INNOVATIONS IN AGRICULTURE EXTENSION

The Editors

Dr. Angelina Patro is currently working as Guest Faculty in the Department of Agricultural Extension Education at Odisha University of Agriculture and Technology. She has obtained her B.Sc Agriculture degree in 2015 from Sam Higginbottom University of Agriculture and Technology, Allahabad, Uttar Pradesh. She passed her M.Sc Ag (Extension Education) from Assam Agricultural University, Jorhat, Assam as a National Talent Scholar (NTS) in 2018 and her Ph.D (Extension Education) in 2023 from College of Agriculture, Odisha University of Agriculture and Technology. She has qualified ICAR-ASRB NET in 2021. So far, she has published thirteen research papers in NAAS, Scopus and Web of Science listed journals, four book chapters and four popular articles. She has also reviewed many research papers and book chapters for various national and international journals.

Dr. Smitashree Das is currently working as Guest Faculty in the Department of Agricultural Extension Education at Odisha University of Agriculture and Technology. She has obtained her B.Sc Agriculture in 2016, M.Sc Agriculture in 2018 and Ph.D Extension Education in 2023 from College of Agriculture, Odisha University of Agriculture and Technology. She has qualified ICAR-ASRB NET in 2018 and UGC-NET IN 2019. She has published several research publications in both National and International Journals. She has been awarded with Ganga Singh Chauhan Memorial Award for academic achievement in her Post Graduation career from the Society of Extension Education, Agra. She has also published a practical manual on Evaluation Scales in Extension Education.

Biswajit Mallick is currently pursuing Doctoral research in the Department of Agricultural Extension Education, Odisha University of Agriculture and Technology (Rank-3). He has qualified ICAR NET-2023 and UGC NET-2022. He has completed his Graduation from Centurion University of Technology and Management, Odisha. He joined Dr. Rajendra Prasad Central Agricultural University, Pusa, Bihar for Post Graduation through ICAR entrance exam as a Junior Research Fellow

INNOVATIONS IN AGRICULTURE EXTENSION

– Editors –

Dr. Angelina Patro

Dr. Smitashree Das

Biswajit Mallick

Department of Agricultural Extension Education
Odisha University of Agriculture and Technology
Bhubaneswar, Odisha

ASTRAL
Publishing Authors Across the Globe

2025

Daya Publishing House®
A Division of
Astral International Pvt. Ltd.
New Delhi – 110 002

ISBN 978-93-5919-940-5

Published by : **Daya Publishing House®**
A Division of
Astral International Pvt. Ltd.
– ISO 9001:2015 Certified Company –
4736/23, Ground Floor, Ansari Road,
Darya Ganj, New Delhi-110 002
Ph. 011-43549197, 8130496929
E-mail: info@astralint.com
Website: www.astralint.com

Printed at : **Replika Press Pvt Ltd**

Preface

Agricultural Extension is not just another discipline; rather it is the only discipline that spans equally across classrooms, research labs and the farms and families of the rural people. The importance of Agricultural Extension is immense, the scope is huge and the possibilities are tremendous. So right now, things can be better, hence it is the time to make Agricultural Extension from theory to practice and from learning to leading.

In an era defined by rapid technological advancement and evolving educational paradigms, the concept of next-generation extension emerges as a beacon of innovation in educational outreach. This preface embarks on a journey into the realm of pioneering practices reshaping the landscape of education. The Next-Gen Extension transcends traditional boundaries, embracing cutting-edge methodologies and technologies to engage learners beyond the confines of conventional classrooms. It embodies a forward-looking approach, harnessing the power of digital platforms, immersive experiences of communication skills, and community-driven initiatives to democratize access to knowledge.

Through the collection of book chapters in this editorial book, we delve into the transformative potential of Next-Gen Extension, pragmatic insights and theoretical frameworks that illuminate its impact on educational outreach. As we navigate this dynamic terrain, we are invited to envision a future where learning knows no bounds, where innovation fosters inclusivity and where Next-Gen Extension serves as a catalyst for positive change in education on a global scale. As we embark on this exploration of pioneering practices in educational outreach, we are reminded of the profound impact that innovation can have on shaping the future of education and society.

The purpose of this Special Issue of the Editorial book ***Innovations in Agriculture Extension*** is to lay the foundation for the next generation of Agricultural Extension knowledge products. This strategy is designed to facilitate a transition from the current, primarily supply-driven educational approach towards a more demand-driven and need-based educational approach that would address the actual needs of the learners in the modern era.

This Editorial book is a basic guide for undergraduate, post-graduate and also for doctorate students which will not only help enhance their learning and formal education but also provide an opportunity for experiencing real-world and lifelong skill development. This unique and innovative learning creation and compilation unfolds new pathways for knowledge dissemination, empowering individuals to thrive in an ever-evolving digital landscape.

This Editorial Book is not free of limitations. We tried our best to make the book free of errors. The content of the book is clear, brief, specific and relevant. Every attempt is made to make this book user-friendly.

Happy Reading!!

Dr. Angelina Patro

Dr. Smitashree Das

Biswajit Mallick

Contents

Chapter 1

Role of Communication in Conflict Resolution

Shruti Pragyan Dash and Saluja Prusty*

Research Scholar, Department of Agricultural Extension Education, Odisha University of Agriculture and Technology, Bhubaneswar 751003, Odisha

ABSTRACT

Communication plays a multifaceted role in conflict resolution demonstrating its effect on managing emotions, fostering understanding and facilitating compromise. Major grounds of conflict in an organization includes communication barriers, allocation of resources, competition and differences in goals or objectives. Conflict resolution strategies in relation to effective communication includes active listening, understanding perspectives, building trust and encouraging open dialogue. Furthermore, communication challenges in conflict resolution are defensiveness, past grudges or resentment. Overall, effective communication is the key to reconciliation and conflict resolution by fostering understanding, facilitating negotiation and validating mediation.

Keywords: *Communication, Conflict, Active-listening, Problem solving, Negotiation, Reconciliation.*

Introduction

Conflict in workplace is inevitable but dealing it in a fair, sensitive and constructive manner is crucial in the process of problem solving and conflict resolution. Effective communication is the cornerstone in resolving conflicts and navigating the problems as it has a multifaceted impact on every individual. Communication serves as a foundation structure in resolving conflicts by fostering empathy and open dialogue through collaboration and compromise. Effective communication not only bridges the conflict of interest but also facilitates negotiation, reconciliation and mediation. Broken communication often leads to escalation of conflict and false assumptions that leads to further dispute among the parties in workplace and to avoid this, active listening and empathetic involvement transcends the barrier, unraveling the complex viewpoints of the conflicting parties to pave the way for reconciliation and resolution.

What is Conflict?

Conflict is disagreement or difference of opinion between individuals, groups or entities in terms of interests, beliefs, or values. Conflict in an organization may arise due to various reasons such as ineffective communication, allocation of resources or contradictory view points between the employees or the employees and the management leading to hindrance in progress of work.

Types of Conflict in an Organization

- **Interpersonal conflict:** Interpersonal conflict arises between individuals due to difference in personalities, opinions or views.
- **Intrapersonal conflict:** Interpersonal conflict arises due to internal conflicts and struggles within an individual in an organization.
- **Intergroup conflict:** Intergroup conflict arises between different groups, teams or departments mostly due to competition for resources within an organization.
- **Intragroup conflict:** Intragroup conflict arises between the members of a group or team due to division of tasks, responsibilities and decision-making process.
- **Role conflict:** Role conflict arises when an individual in an organization is expected to serve multiple roles simultaneously with inadequate time and resource.
- **Vertical conflict:** Vertical conflict occurs between individuals or groups at different hierarchical levels, say between employees and management.
- **Horizontal conflict:** Horizontal conflict occurs between individuals or groups at same hierarchical levels, say between two teams or departments having similar power and resources.
- **Cultural conflict:** Culture conflict arises due to difference in backgrounds, cultural norms,ethics or values of the employees.

Reasons of Conflict in an Organization

- **Communication barriers:** Ineffective communication can lead to misunderstandings which generates conflict.
- **Misalignment of goals or objectives**: Differences in goals or objectives among individuals in a team or departments.
- **Competition:** Competition for power, resources and recognition may create conflict especially in places where individual achievement is more recognised than teamwork.
- **Personality clashes:** Difference in personalities, way of doing tasks and decision-making abilities can lead to conflict.
- **Role ambiguity:** Unclear roles and responsibilities in working environment can lead to confusion and conflict, especially when an employee feel overworked or undervalued.
- **Resource allocation:** Issues related to budget, staffing and division of resources can lead to conflict.
- **Policy change:** Changes or uncertainties in institutional policies can create tension and conflict among the employees.

Conflict Management Strategies

- **Intercommunication:** clear and mutual exchange of conversation can help address misunderstanding.
- **Collaboration:** building teamwork that helps to find mutually beneficial solution to problems which strengths relationships in workplace.
- **Assertiveness:** clearly stating one's needs and boundaries while respecting those of others can help prevent conflict.
- **Compromise:** Mutual concession between both the parties that give up some part of their demands to reach a middle ground.
- **Avoidance:** It is better to step down from a heated situation temporarily and address the problem in a cool environment later.
- **Problem solving:** Address the root cause of a problem using analytical skills and finding creative solutions can alleviate the conflict.
- **Accommodation:** Considering the need of others before one's own need to maintain harmony but not at the expense of own well-being.
- **Mediation:** Taking opinion from a neutral third-party in decision-making process can help find a resolution to the conflict.

Role of Communication in Conflict Resolution

Effective communication is most crucial for conflict resolution as it serves as a bridge between conflicting parties by fostering empathy and mutual respect and facilitating understanding between them. Communication helps in clarifying misunderstandings by addressing the root cause of the problem, identifying primary issues and discovering Potential solutions. Clear and open communication creates a conducive environment where individuals can express their perspectives, concerns and emotions. Ultimately, effective communication is necessary for alleviating conflicts, promoting reconciliation and fostering long term relationship built in terms of understanding.

Strategies for Effective Communication in Conflict Resolution

Active Listening

Active listening plays a crucial role in conflict resolution by fostering understanding, empathy, and concerns of every individual involved in the conflicting situation. It involves clear concentrating on what is being said, understanding the message, and reflecting back or responding thoughtfully. Through active listening, conflicts can be alleviated as each individual feels heard and understood, leading to effective communication,problem-solving and resolution of the ongoing conflict.

Roles of Active Listening as Part of Effective Communication in Conflict Resolution

- ✰ **Perspective-taking:** Active listening allows an individual to comprehend perspective of everyone involved in the conflict and also considers their viewpoint,emotions and concerns.
- ✰ **Clarifying misunderstandings:** Active listening Plays a vital role in clarifying misunderstandings as it involves questioning, paraphrasing or summarizing to ensure everyone's understanding is on the same page.
- ✰ **Building trust:** Active listening helps build trust by creating a conducive environment for the conflicting parties by creating genuine concern and interest.
- ✰ **Promoting open dialogue:** Encouraging constructive dialogue by promoting open and honest conversation that helps in problem solving and conflict resolution.
- ✰ **Reducing defensiveness:** Active listening helps reducing defensiveness by allowing more effective exchange of ideas and perspectives among the individuals.
- ✰ **Validation of feelings:** Active listening helps in acknowledging the feelings and emotions of others which helps foster empathy and understanding among the individuals.
- ✰ **Facilitating agreement:** Active listening contributes to developing mutually acceptable solutions by identification of common ground leading to effective resolution of conflict.

- **Empowering individuals:** The feeling of heard and understood within an individual enhances the urge and willingness to engage in the conflict resolution process and providing beneficial solutions.

Non-violent Communication Techniques

- **Observation:** Focusing on actions and behaviour of individuals than interpretation and describing the situation without evaluation.
- **Feelings:** Emphasizing emotions rather than blame or criticism and expression how an individual feel about a particular situation.
- **Needs:** Identifying the underlying needs or values of an individual that contributes to their feelings.
- **Requests:** Make specific, genuine requests that are feasible and concrete.
- **Empathy:** Acknowledging the other person's feelings and needs by showing empathy even in case of disagreement.
- **Problem-solving:** Collaborate with the other person or team to find mutually beneficial solutions that benefits the needs of both parties.

Communication Challenges in Conflict Resolution

- **Defensiveness:** The feeling of being attacked may develop defensiveness within an individual making it difficult to listen and empathize with the other person's views.
- **Anger and Hostility:** Strong emotional response like anger can hinder clear communication, leading to further escalation of the conflict.
- **Fear:** Fear of being misunderstood or confrontation can prevent from the members from communicating openly and honestly.
- **Resentment:** Past grudges or resentment can deter one's focus on resolving the current conflict.
- **Lack of Empathy:** Inability to empathize with the other person's perspectives and feelings can lead to misunderstandings and broken communication.
- **Communication Style:** Different communication styles, such as avoidance or passive-aggressive behaviour can hamper the process of problem solving.

Conclusion

Effective communication plays a crucial role in conflict resolution as it facilitates understanding by fostering empathy and concern through collaboration and compromise. Active listening facilitates open dialogue and promotes transparency that helps in exchange of ideas and viewpoints which further leads to de-escalation of conflict. Furthermore, clear and open conversation created a conducive environment where each and every individual can share their perspectives, concerns and viewpoints that helps in problem solving and conflict resolution.

REFERENCES

Adejimola, A. S. (2009). Language and communication in conflict resolution. Journal of law and conflict resolution, 1(1), 001-009.

Behfar, K. J., Peterson, R. S., Mannix, E. A., and Trochim, W. M. (2008). The critical role of conflict resolution in teams: A close look at the links between conflict type, conflict management strategies, and team outcomes. Journal of applied psychology, 93(1), 170.

Choudrie, J. (2005). Understanding the role of communication and conflict on reengineering team development. Journal of Enterprise Information Management, 18(1), 64-78.

Deutsch, M., Coleman, P. T., and Marcus, E. C. (Eds.). (2011). The handbook of conflict resolution: Theory and practice. John Wiley and Sons.

Mohr, J., and Spekman, R. (1994). Characteristics of partnership success: partnership attributes, communication behavior, and conflict resolution techniques. Strategic management journal, 15(2), 135-152.

Neiman, T. S. (2011). Communication in Conflict and Problem-Solving: A Study of Dialogue in Everyday Life.

Olekalns, M., Putnam, L. L., Weingart, L. R., and Metcalf, L. (2007). Communication processes and conflict management. In The psychology of conflict and conflict management in organizations (pp. 99-132). Psychology Press.

Ott, M. C. (1972). Mediation as a method of conflict resolution: Two cases. International Organization, 26(4), 595-618.

Chapter 2

Social Media Affecting Dietary Habits

Rosalin Sahoo and Sipra Baral

MSc Scholar, Food and Nutrition, College of Community Science, Odisha University of Agriculture and Technology, Bhubaneswar 751003, Odisha

ABSTRACT

Social media is a cutting-edge technology that enables online connections and information sharing between users. Over the past ten years, social media has played a major role in reconstructing the social landscape of societies. Social media is a crucial tool for raising standards of living and society in the twenty-first century. Social media provides a platform for people to share, create, comment on, and exchange content with others in various networks. Additionally, social media is essential for improving an individual's nutritional status and food quality. Social media tools, such as blogs, social networks, and media-sharing websites, assist practitioners in nutrition and dietetics in connecting with the public and reaching a wider audience. Dietetics practice has been changed by social media in many ways and has opened up new channels for disseminating knowledge about food and nutrition. Peer-to-peer support, public health campaigns, patient education, and virtual nutrition counseling have all benefited from the use of social media. Content creator of nutrition and dietetics are using social media to network and work together with colleagues, look for a job, stay in trend with research, support causes, advertise goods and services, and launch a business. Nutritionists and dietitians must understand the professional benefits and drawbacks of social media in the current digital era. Inappropriate use of social media have a negative impact on both the profession and the individual practitioner. Teenagers and young adults learn about nutrition and maintain healthy diets through social media.

Keywords: *Adolescent, Food craving, Nutrition, Social change, Social media.*

Introduction

Young adults (ages 18–25) and adolescents (ages 10–19) find it difficult to maintain healthy diets. More than 80 per cent of teenagers, according to national health surveys, do not follow the recommendations for a healthy diet. When young people start living on their own in their early adult years, dietary habits also frequently get worse. Increased risks for diabetes, cardiovascular disease, and other chronic illnesses may result from this over time. Adolescent and young adult interventions, which are frequently carried out in high school and colleges, have had some success with behaviour-focused education, assessments that provide feedback, and peer involvement. As the use of social media is increasing day by day, teenagers and young adults are accessing it for their food habits. Innovative computing and mobile-based health and nutrition platforms take advantage of young people's through mobile devices to offer nutrition-related knowledge, to make dietary counseling and clinical communication easier, and promote behaviour tracking and self-evaluations.

Social Media and Food Craving

Social media suggests which foods to be consumed in moderate or high and also provides us ideas to make a healthy meal. Most of the people lose control of their ability for eating and are more likely to engage in binge eating. Eating more foods in a short amount of time experiencing a loss of control is called as binge eating. The potential detrimental impacts of social media may rise with the amount of time spent in front of devices. Excessive use of internet, alters behaviour and is associated with impulsivity and a loss of self-control, leading to eating disorders in adolescent. Adolescent girls who used Instagram on a daily basis exhibited noticeably more disordered eating behaviours. Furthermore, the exposure of screens to blue light at night alters the way that glucose is metabolized, potentially contributing to adult obesity. Social media has given food and beverage companies–who often employ media through influencer advertising, in which celebrities collaborate with firms to endorse and promote products to their followers–new marketing opportunities. Several viewer behaviors, including product consumption, the urge to like and share postings, and brand recognition, have been connected to influencer marketing of food and beverage items.

14 per cent of respondents said they were craving the product after seeing it advertised, and 8 per cent said they had bought one. Adolescents frequently interact with brands of unhealthy food on social media. As a result, exposure to fatty and sugary foods rises. Lastly, it's critical to remember that social media has been linked to impulsive behaviour and food cravings, both of which may overlap and exacerbate issues with eating behaviour.

Pros of Social Media

Despite all the gloomy articles about how social media harms us, following your favourite accounts on the platform can have advantages.

Uplifts and Motivates

Following healthy role models can give you the confidence to make great changes in your life, even though there's a narrow line dividing aspirational from simply unrealistic social media posts. Think considering subscribing to people who, although open about their own issues, have diets or fitness levels that you find motivational.

Skills and Information

Do you know how to cut a jackfruit properly ? There's a YouTube video to guide. Do you need a high fibre meal plan for a month? Pinterest and Instagram is on our side. Using social media to get information that is appropriate for your own particular dietary needs has never been simpler.

Fosters a Healthier Food Community

One of social media's best qualities is its ability to bring people together who have similar interests. Whether you're a lover of tropical fruits or this is your first effort at a low-carb diet, it's easy to connect with others online who are going through similar things. After you've met "your people," you could also find it easier to stick to your strategy for better health.

Open up to New Content

Gone are the days when trying a new recipe meant getting through the cookbooks. With just a few clicks on social media these days, we may be introduced to new recipes, ingredients, and even perspectives on food. If you follow chefs from around the globe, you might get motivated to look for an ingredient you've never used. For a more conscious approach to eating, consider the philosophy of an Intuitive Eating dietitian.

Cons of Social Media

Social media can be beneficial, but it also has some negative aspects. When utilizing different platforms, keep these possible drawbacks in mind.

Distracted Eating Leads to Overconsumption

The act of scrolling through social media can have an adverse effect on eating habits rather than just the content itself. Overeating is typically the result of distracted eating. People tended to eat more when they weren't paying attention to their meals, according to a 2013 meta-analysis of 24 studies. In a different 2020 study, people who are distracted while eating eat less, and also enjoy their food less. It's usually a good to off your electronics like computers, tablets, and phones while eating.

Creates Unrealistic Beliefs and Expectations

To tell the truth, social media doesn't really set realistic standards. Influencers and regular people can showcase their good points–specifically, their eating habits–on the internet. We will eventually be disillusioned with our own diets if we subject ourselves to these purported standards of perfection. When anything on social media makes you feel horrible about yourself, try to remember that you are merely witnessing a glimpse of someone else's life. For many, social media is a full-time job in the interim, but it's easy to forget that. Following someone who cooks vegan or gluten-free food may be challenging if you have a demanding work and finicky eaters in your family, for example. Studies reveal that women often edit their images with filters or retouch them before posting them on social media. Furthermore, some data suggests that young females who spend more time photoshopping images and changing their online personas on social media are more likely to suffer from eating disorders and have issues with body image. Furthermore, studies have suggested a possible association between the use of certain photo-editing apps and social media sites and increased acceptance of cosmetic surgery and decreased self-esteem.

Ads Influence Us More

Despite our tendency to believe differently, advertisements are used by marketers because they work. Additionally, because social media sites now display customized ads based on your personal information, advertising is becoming more and more individualized. According to research published in 2016, this could have an effect on our purchasing habits, behaviors, and self-perceptions. It is clear that social media ads have a big influence on behavior overall and can make us seek for items we wouldn't normally, even if more research is needed to completely understand the relationship between these ads and dietary habits.

Eating Disorders

Numerous studies have demonstrated the negative effects of social media on our attitudes toward food, particularly for younger generations. A big 2016 research of young adults found a "strong and consistent association" between using social media and having anxiety about what to eat. Another 2020 study found a connection between children's usage of social media and disordered eating behaviors and thoughts in the seventh and eighth grades. Social media issues including idealizing unreachable ideal, comparing oneself to peers, and hearing hurtful comments about one's weight can make disordered eating worse.

Conclusion

Researchers are worried that food-related content on social media platforms is influencing. Content creators or food vloggers on social media showcase varieties of foods to maintain themselves in this competitive market. Content creators may progressively change their content to be unhealthier in order to remain competitive, given that unhealthy foods have greater engagement and reach metrics. We can all influence people online, so it's not just the food industry's sponsored spots that are to blame. One study found that children and teenagers are exposed to thirty to eighty-nine times a week on social media applications with food marketing, with fast food and sugary drinks being the most frequent categories.

REFERENCES

Chung A, Vieira D, Donley T, Tan N, Jean-Louis G, Kiely Gouley K, Seixas A. Adolescent Peer Influence on Eating Behaviors via Social Media: Scoping Review. J Med Internet Res. 2021 Jun 3;23(6):e19697. doi: 10.2196/19697. PMID: 34081018; PMCID: PMC8212626.

Fortin B, Yazbeck M, Peer effects, fast food consumption and adolescent weight gain, *J. Health Econ.* 42 (July) (2015) 125–138.

Niemeier HM, Raynor HA, Lloyd-Richardson EE, Rogers ML, Wing RR, Fast food consumption and breakfast skipping: predictors of weight gain from adolescence to adulthood in a nationally representative sample, *J. Adolesc. Health* 39 (December (6)) (2006) 842–849.

Mikkilä V, Räsänen L, Raitakari OT, Pietinen P, Viikari J, Consistent dietary patterns identified from childhood to adulthood: the cardiovascular risk in young finns study, *Br. J. Nutr.* 93 (June (6)) (2005) 923–931.

Retrived from https://www.verywellfit.com/how-does-social-media-affect-eating-habits-5188578 on 19/03/24

Retrived from https://www.jandonline.org/article/S2212-2672(16)31071-1/abstract#secsectitle0110 on 19/03/24

Retrived from https://www.ncbi.nlm.nih.gov/pmc/articles/PMC6983924/ on 19/03/24

Retrived from https://jeatdisord.biomedcentral.com/articles/10.1186/s40337-022-00698-4 on 19/03/24

Chapter 3

Mobile Phone Application for Agricultural Extension in India

Manisa Hota

Research Scholar, Department of Agricultural Extension, Odisha University of Agriculture and Technology, Bhubaneswar 751003, Odisha
e-mail: mhota203@gmail.com

ABSTRACT

Millions of farmers rely mostly on agriculture for their livelihood, making it most significant sector of the Indian economy. The availability of land and water resources is nearly depleted, commodity prices are constantly fluctuating, most marginal and small farmers make very little money, and most importantly it is difficult to obtain information. Soft resources including knowledge and skills are just as important as physical resources in modern agriculture. However, estimates suggest that 60 per cent of farmers lack access to any information source regarding cutting edge agricultural technologies, creating a massive adoption gap. Nearly everyone in the world has a cell phone these days. We have nothing to lose by exploiting this enormous reach to transmit agricultural knowledge in multimodal form in developing nations like India, where it has the potential to completely transform agriculture. Due to growing reliance, mobile phones are spreading throughout the globe as a standard means of communication, particularly in the agriculture sector.

Keywords: *Agriculture extension, Phone, Application, Farmers, Services.*

Introduction

We know agriculture is one of the important sectors of the Indian economy and since the time of independence it has been the major contributor to the GDP of the country. In case of increasing food production research, extension and farmers effort play the major role. As per the data concerned the food production in 1950-51 which was 50 million tonnes increased to 329.68 million tonnes in 2022-23 (MoA, GOI, 2023). Again, considering the demand of increased population of our country there is a need in further increment of food production. In previous time, technology and agriculture were considered as two opposite ends of value chain system as we were more biased towards conventional method. But today we have so many evidences of use of technologies in our farming system and mobile phone application is one of them. Now a days farmers who are not using the modern technology like mobile apps are not considered as progressive farmers. Problems like climate issues, soil erosion, mineral depletion, downgrading water level can be solved by linking agriculture and technology. So to manage the farms of the farmers effectively and efficiently mobile applications are increasingly used.

Agriculture Applications for Farmers

- **IFFCO Kisan Agriculture App:** This app was launched in 2015 and it is a developed by Indian farmers' fertilizer cooperative Ltd. The main role of this app is to help the Indian farmers in decision making through modified information which are related to their needs. Its goal is to provide Indian farmers with tailored information that will enable them to make well informed decisions. Additionally during the profiling stage the user has access to a range of informational modules in a form of text, image, audio and videos in the chosen language such as agricultural advisory, weather, market pricing and farm information library. Additionally, the app provides phone number for kisan call center services.
- **Krishi Jagran:** Krishi jagran is an app that offers everything from crop calenders to cultivation guidelines, trending agri news, subsidies a career in agriculture, and comprehensive information on crop security, pest, and disease management.Farmers may connect with the rest of the world and get all the pertinent information related to agriculture on our user-friendly agriplatform.
- **Pusa Krishi:** the union agriculture ministry introduced the government app in 2016 with the goal of assisting farmers in learning about technology created by the Indian agriculture research institute (IARI) that will help them receive higher returns. The application furnishes farmers with data pertaining to novel crop varieties created by the Indian council of agricultural research (ICAR), cultivation techniques

that conserve resources and farm equipment, the utilization of which can enhance farmers' profits.

- ☆ **Agri App:** The Agri App is a comprehensive tool designed with farmers in mind, including comprehensive data on crop protection, crop production, and other related agricultural services. Additionally,it gives farmers access to all data about crop varieties,soil and climate, harvesting techniques, and storage methods associated with crops in the "High value, low product category. This app also offers the ability to speak with professionals, video based learning, the most recent news, and online markets for things like insecticides and fertilizers.
- ☆ **Bijak App:** An online marketplace Bijak links reputable farmers, loaders, and commission agents throughout India. It guarantees end to end services including advance payment guarantee, limit facility, secured payments, mandi rates *etc.* to suppliers and purchasers (commission agents).
- ☆ **Crop insurance:** This amazing tool gives farmers information on deadlines, corporate contacts, insurance rates for various crops. It also helps them figure out how much to pay for their harvest. For farmers, it serves as a calculator and a reminder regarding their insurance. Additionally it can be used to obtain information on any notified crop in any notified area,including facts about the standard sum insured, extended sum insured, premium details, subsidy information. It is also connected to its website, which serves all interested parties, including banks, insurance providers, states and farmers.
- ☆ **Agri-Market:** The software was created to inform farmers about crop prices and dissuade them from participating in distress sales. Using the Agrimarket mobile app, farmers can obtain information about crop prices in markets within 50 kilometres of their device's location.
- ☆ **Kheti-Badi:** A social application, aims to promote and support organic farming while providing essential information to Indian farmers. It facilitates the transition from conventional chemical farming to organic practices. Presently, the app supports four languages: Hindi, English, Marathi, and Gujarati.
- ☆ **Shetkari Mitra:** Indian farmers can use the multipurpose smartphone app shektari mitra. It offers details and expertise on government programmes, crop management, agribusiness policies and guidelines, market prices, and agricultural success stories.
- ☆ **Kishan Suvidha:** It was started in 2016 with the goal of empowering farmers and promoting rural development by prime minister Narendra Modi. The app has a clean, easy to use layout and includes useful features including current and forecasted weather for the next five days, market prices for crops and commodities in the closest town, and

information on fertilisers, seeds, machines and other related topics. The app is more widely available because it may be used in multiple languages.

- **Kisan call centers:** These call centers aim to provide farmers with prompt, locally spoken responses to their inquires. Every state has call centers that are meant to handle calls from all cross the nation. These call centers handle inquiries about the agricultural and associated industries. Information and communication technology (ICT) and agricultural technology two formerly distinct fields with distinct work cultures and specialized areas, are combined to produce the kisan call centre.
- **mKisan**: The mFarmer initiative challenge fund helped to launch the mKisan projects is being carried out by the international livestock research institute (ILRI), India has partnered with Handygo Technologies, a mobile value-added service provider, along with CABI South Asia and Digital Green, an NGO dedicated to promoting video-enabled extension services. The project's goal is to provide small holders in India with a full range of agro advisory services. The project aims to achieve several goals, including delivering daily updates on agrometeorology, crop pest and livestock disease outbreaks, market information, local service providers, and enhancing information access for women farmers. Information that has been verified by science and is useful will be made available by the CABI through its Direct2farm Repository. The mKisan initiative seeks to connect with one milion small holders.

Essential Components for the Mobile Agricultural Advisory App's Success

- **Partnership:** collaboration with relevant content producers and software vendors will be essential to the mobile based agricultural consulting services' success.
- **Content:** The farming community benefits more from location and farmer specific material than from generalized content. Furthermore, farmers greatly value and demand extra information, such as the availability of pesticides and fertilizers in the local area.
- **Innovation:** To guarantee the survival of the services, competent mobile application development and ongoing innovation are required. The 60–80-character SMS limit for agricultural information has been surpassed, and voice, images, and videos are now more in demand. With mobile phones becoming more affordable every day, developing apps for smart phones and medium range phones can significantly boost the efficacy and popularity of the business.

- ☆ **Timeliness:** A major component influencing agricultural information is timeliness. For farmers, communications that arrive too late or too early are of little use, and service eventually loses confidence.
- ☆ **Quality and credibility:** The success of the service is largely dependent on the quality of the message sent to the farmer, regardless of the medium. Weather, disease, and pest assault forecasts are far more sought than preventions, real time commodity market prices and price projections, and personalized and location specific messages are more useful than generic ones. The longevity of the service depends on the clients' perception of the service providers' legitimacy.

Conclusion

Although there are many advantages to using mobile applications in india, there are also many drawbacks. These include a dearth of locally relevant and mobile friendly digital content, issues with electricity and network in rural areas, low literacy rates, and a multitude of regional and tribal dialects. Furthermore, a lot of pilot programs are started in project mode, and the biggest challenge after the project term is sustainability. Therefore, mobile applications must be incorporated with ongoing agricultural extension programs and techniques rather than being isolated projects. To meet the information and knowledge needs of farmers, public private partnerships for project implementation and web to mobile, voice, image and text integrated applications must be developed. It will be encouraged for farmers to use mobile-based advisory services. A multimedia format containing geographical and farmer specific information must be created based on an assessment of farmers' needs before it can be distributed to the millions of farm families.

REFERENCES

Belakeri, P., Prasad, C. K., Bajantri, S., Mahantesh, M. T., Maruthi, S. T., and Rudresh, G. N. (2017). Trends of mobile applications in farming. *International Journal of Current Microbiology and Applied Sciences, 6*(7), 2499-2512. https://krishijagran.com/agripedia/top-10-agricultural-mobile-apps-for-farmers-in-2021/

Engotoit, B., Moya, M. B., Mayoka, K. G., and Abima, B. (2016). A Mobile-Based Communication Adoption Model for agricultural market information dissemination in Uganda.

Khan, N., Siddiqui, B. N., Khan, N., Khan, F., Ullah, N., Ihtisham, M.,. and Muhammad, S. (2020). Analyzing mobile phone usage in agricultural modernization and rural development. *International Journal of Agricultural Extension, 8*(2), 139-147.

Nicholase-Ere, O. (2017). Dissemination of agricultural information to farmers using ICT. *International Journal of Computer Applications, 179*(7), 27-31.

Patel, M. R., Patel, M. V., and Patel, R. A. (2017). Assessment of Kisan Mobile Advisory (KMA) Service for Dissemination of Agricultural Information in Mehsana District of Gujarat State. *Indian Journal of Extension Education, 53*(1), 90-93.

Qiang, C. Z., Kuek, S. C., Dymond, A., and Esselaar, S. (2012). Mobile applications for agriculture and rural development.

Saravanan, R., and Bhattacharjee, S. (2014). Mobile phone applications for agricultural extension in India. *Mobile phones for agricultural extension: Worldwide mAgri Innovations and Promise for Future.* (*Edited by Saravanan, R*), 1-75.

Chapter 4

Exploring Communication Styles in Leadership Roles

Saluja Prusty[1], Shruti Pragyan Dash[2] and Jyotishree Anshuman[3]*

[1,2]Msc Scholar, [3]PhD Scholar, Department of Agricultural Extension Education, Odisha University of Agriculture and Technology, Bhubaneswar 751003, Odisha

ABSTRACT

Leadership encompasses the diverse tasks of influencing, inspiring, and guiding others toward a shared objective. Effective leadership hinges on dynamic communication, which fosters collaboration, understanding, and motivation among leaders, followers, and external stakeholders. This chapter explores the integral relationship between different leadership styles and communication styles, emphasizing the significance of developing excellent communication skills for successful leadership. Clear and compelling communication enables leaders to articulate their vision, convey urgency, and inspire enthusiasm, ultimately driving collective action towards shared objectives.

Keywords: *Communication, Communication styles, Leadership, Leadership styles.*

Introduction

Leadership entails influencing, inspiring, and guiding others to accomplish a goal. It's a relational, dynamic, and practical skill that can be acquired and utilized in diverse contexts. Leadership requires collaboration and interaction among leaders, followers, and sometimes outside team/group/organisation to work for accomplishing the common goal. The power of leadership is the power of integrating. Communication helps people integrate by building relationships, expressing feelings and sharing experiences and needs. Communication also helps people understand others, conquer differences, and clarify situations. Effective leadership relies heavily on developing exceptional communication skills. Leaders must articulate knowledge and ideas effectively to instill a sense of urgency and enthusiasm in others. If a leader fails to convey a message clearly and inspire others to take action, then the message becomes irrelevant.

Communication

Communication, derived from Latin, signifies "to share." The word communication means sharing the same thoughts, conceptions, and observations. In essence, communication involves the exchange and engagement of ideas, facts, opinions, feelings, or attitudes. Communication serves as the essence of management and leadership. The fundamental functions of management–planning, staffing, supervision, and organization–rely heavily on effective communication. It's a reciprocal process of conveying information or messages between individuals or groups. The process involves at least one sender and receiver to exchange messages. These messages may consist of ideas, thoughts, imaginations, or emotions.

Communication Styles

Communication styles are how people express themselves and respond to others in conversation and interaction in day-to-day life. Communication styles encompass the diverse approaches individuals take in communicating. They help individuals become aware and optimize their communication methods. These styles are categorized according to an individual's preferred manner of expressing ideas, thoughts, and emotions.

The 5 Basic Communication Styles, Division by Bourne (1995

Psychologist Edmund J. Bourne identified five distinct types of communication styles.

1. **Submissive:** The submissive communication style prioritizes satisfying others while evading conflicts. Submissive communicators often perceive others as having more rights and greater contributions to conversations, meetings, and discussions. They prioritize the needs of others over their own.
2. **Aggressive:** In contrast to submissive communicators, aggressive communicators prioritize their own needs over others', often seeking to win at any cost, even if it means disregarding the needs of others. They perceive their opinions as paramount in discussions and assert a greater entitlement to contribute than others. Consequently, the significance of the information they convey may be overshadowed by their insensitive delivery.

3. **Passive-aggressive:** The passive-aggressive communication style involves individuals appearing passive on the surface but expressing their anger indirectly. They often behave in this manner due to unresolved issues or unfavorable circumstances. To express their frustration and dissatisfaction, passive-aggressive communicators subtly undermine the person they are communicating with, even if it means exacerbating their own situation.
4. **Manipulative:** The manipulative communication style entails using cunning behavior to achieve desired outcomes. Manipulative communicators employ trickery and calculation to influence and control others to fulfil their own agenda. The thoughts, ideas, and opinions they convey may carry hidden meanings, making others initially unaware of their ulterior motives.
5. **Assertive:** The assertive communication style stems from self-esteem and represents the healthiest and most effective approach to communication. Assertive individuals neither lean too far toward passivity nor aggression. They communicate directly and effectively, avoiding passive-aggressiveness or manipulation. Additionally, assertive communicators respect both their own and others' rights and needs equally.

Communication Styles, Division by Alessandra and Hunsaker (1993)

Alessandra and Hunsaker delineate four communication styles: Realtors, Socializers, Thinkers, and Directors, categorized along two dimensions: Direct versus indirect styles of communication.

1. **Direct style of communication:** The direct communicators tend to be assertive and express their opinions openly. They radiate self confidence, maintain eye contact and dominate meetings with their outspoken communication.
2. **Indirect style of communication:** They typically exhibit patience, diplomacy, and cooperation in various situations. In social gatherings, they are less inclined to take the lead and may hesitate to contribute in meetings because of their more tentative communication style.
3. **Supporting style of communication:** Supportive communicators, also referred to as people-oriented communicators, establish closer connections with those they communicate with. They freely share their feelings and opinions, expressing emotions such as joy, sadness, and confusion without hesitation. They gravitate towards cultivating relaxed and warmer relationships.
4. **Controlling style of communication:** Controlling style of communication involves task oriented communication. They tend to

share their feelings less, pay less attention to the feelings and opinions of others, excel at time management, prefer working alone, avoid being noisy and violent.

5. **Relator's communication style:** Characterized by an indirect and supportive communication style, they prioritize the feelings of those they communicate with over the efficiency of tasks. They typically steer clear of conflicts and derive satisfaction from personal relationships.
6. **Socializer's communication style:** Characterized by a direct and supportive communication style, their primary motivations lie in receiving acknowledgment, admiration, and compliments from others. They often aim to motivate and positively influence people in a friendly manner.
7. **Thinker's communication style:** They are categorized by a controlling and indirect communication style. Their focus typically revolves around expectations and outcomes, shaping their communication through inquiries about how things should function to ascertain their effectiveness.
8. **Director's communication style:** Characterized by a controlling and direct communication style, these individuals are often perceived as task-oriented leaders who prioritize outcomes. They possess strong leadership skills and exhibit a commanding presence when communicating with others.

Communication Styles, Division by Murphy *et al.*

Mark Murphy and his team have identified four communication styles.

1. **Analytical:** Analytical communicators prioritize data and value individuals who communicate using precise language and enjoy discussions centered around facts and figures. Lack of supporting facts and figures can undermine credibility in the eyes of an analytical communicator.
2. **Personal:** Individuals with a personal communication style prioritize emotions and value the use of emotional language to establish connections with others. They often exhibit diplomatic skills and excel at listening, as they place significant importance on understanding others' thoughts and emotions.
3. **Intuitive:** Intuitive communicators are action-oriented and concentrate on the broader perspective. They prefer broad overviews to meticulously ordered events or explanations of the situation and tend to avoid excessive details.
4. **Functional:** They are process-oriented individuals who prefer detailed discussions over quick, to-the-point conversations favored by intuitive communicators. In contrast, they concentrate on examining details

and processes, taking their time to explore solutions and generate new ideas.

Workplace Communication Styles, Division by Cox

There are 4 distinct types of communication styles.

1. **The action-oriented style:** Action-oriented communicators are motivated to accomplish tasks and deliver positive outcomes. They embrace change and are dedicated to translating their plans into actions to enhance their present circumstances.
2. **The process-oriented style:** Their goal is to reach logical solutions, employing rational arguments rather than forceful tactics to attain desired outcomes.
3. **The people-oriented style:** These communicators are attuned to the needs of others and make a deliberate effort to uphold the rights of all participants in conversations. They are empathetic, assertive individuals whose core values revolve around teamwork and a strong ethical compass.
4. **The idea-oriented style:** These individuals perceive the world through models, concepts, theories, and abstractions. They constantly seek new possibilities and adopt a future-oriented approach.

Communication Styles, Division by Portolese Dias

Laura Portolese Dias acknowledges that in addition to organizational communication systems, individuals also rely on their personal communication styles. She identifies four distinct styles, which are:

1. **The expresser communication style:** Individuals with an expresser communication style are often enthusiastic, unafraid of challenges, and rely on their intuition. However, relying solely on instinct may pose challenges in situations that necessitate decisions based on factual evidence and concrete data.
2. **The driver communication style:** True to their name, drivers opt for the direct route. They are decisive and guided by firm beliefs and perspectives, leaving little space for small talk or inefficiency.
3. **The relator communication style:** Individuals who employ a relator communication style are apprehensive about rejection and worry that others will dismiss their suggestions and ideas. Consequently, they may become reserved and withdrawn when faced with challenges.
4. **The analytical communication style:** They adopt a methodical mindset, and their logical approach is evident in their communication style. They strive to thoroughly analyze all aspects of a problem, delving into its technical intricacies.

Leadership Styles

Leadership style is a leader's approach to motivating, directing and managing their team and team members. A leader often carries various leadership styles to address different business situations. Different types of leadership styles:

1. **Autocratic:** Also known as the "authoritarian style of leadership," this type of leader prioritizes results and team effectiveness. They usually possess self-confidence and self-motivation, are reliable, and adhere to rules and regulations. Often, they make decisions independently or in consultation with a select few, expecting employees to comply with their directives without question.
2. **Bureaucratic:** Similar to autocratic leaders, they expect strict adherence to rules and procedures from their team members. This style is particularly effective in highly regulated industries or departments such as healthcare, finance, or government. They are detail-oriented and task-focused, possess strong willpower and self-discipline, value rules and structure, and exhibit a strong work ethic.
3. **Coaching:** These leaders establish clear expectations and foster a positive, motivating environment to encourage the growth of each employee. They have a keen ability to identify their team members' strengths, weaknesses, and motivations, enabling them to provide tailored support for improvement. This style is highly beneficial for both employers and the employees they oversee.
4. **Laissez-faire leadership style:** This style contrasts with autocratic leadership, as it emphasizes delegating numerous tasks to team members with minimal supervision. By not micromanaging employees, leaders employing this style often have more time available for other projects.
5. **Democratic:** The democratic style, also known as the "participative style," blends elements of both autocratic and laissez-faire approaches. Leaders employing this style encourage a work environment where everyone is invited to share their ideas. They are characterized by flexibility and rationality in their approach to leadership
6. **Pacesetter:** The pacesetting style is highly effective for swiftly achieving results, especially in situations where team members need to be motivated. Leaders employing this style prioritize performance and frequently establish lofty standards, holding their team members accountable for reaching their objectives.
7. **Servant:** They embody a people-first mindset, recognizing that when team members feel fulfilled both personally and professionally, they are more productive and likely to consistently deliver exceptional

work. Their emphasis on employee satisfaction and collaboration often leads to earning higher levels of respect.

8. **Visionary:** They demonstrate a strong capability to propel progress and initiate change by inspiring employees and gaining trust for new ideas.

Significance of Communication in Leadership

Inadequate communication can result in decreased morale and unmet performance goals, underscoring the importance of strong communication skills for leaders. Effective leaders inspire positive, forward-thinking change by motivating those around them to collaborate towards shared objectives, with communication serving as their most potent tool.

Efficient communication is vital for building trust, instilling confidence, aligning efforts towards common goals, and fostering positive change. Conversely, a lack of communication can result in misunderstandings, strained relationships, and ultimately create barriers that impede progress.

Various communication styles are frequently cited as the primary cause of poor communication, which can escalate to more significant problems like unclear priorities and heightened stress levels. It is crucial to identify a leader's communication style to enhance their understanding of how they interact with and are perceived by employees throughout the organization.

Conclusion

Communication Styles play a vital role in leadership effectiveness. Different leaders may adopt various communication styles based on their personality, situation, and organisational culture. Understanding the communication style is essential for effective interactions in both personal and professional settings. By recognizing communication preferences and tendencies, one can enhance clarity, empathy, and rapport in interactions with others. Whether we learn towards assertive, passive, or passive-aggressive communication styles, self-awareness empowers to adapt and communicate more effectively with diverse personalities. Embracing a nuanced understanding of communication styles fosters healthier relationships, improved conflict resolution, and ultimately, greater success in navigating various social and professional contexts.

REFERENCES

Abbasi, M. H., Siddiqi, A., and Azim, R. U. A. (2011). Role of effective communications for enhancing leadership and entrepreneurial skills in university students. International Journal of Business and Social Science, 2(10).

Charteris-Black, J. (2006). The communication of leadership: The design of leadership style. Routledge.

De Vries, R. E., Bakker-Pieper, A., and Oostenveld, W. (2010). Leadership= communication? The relations of leaders' communication styles with leadership styles, knowledge sharing and leadership outcomes. Journal of business and psychology, 25, 367-380.

Eagly, A. H., and Johnson, B. T. (1990). Gender and leadership style: A meta-analysis. Psychological bulletin, 108(2), 233.

Fairhurst, G. T., and Connaughton, S. L. (2014). Leadership communication. The SAGE handbook of organizational communication: Advances in theory, research, and method, 401-423.

Hackman, M. Z., and Johnson, C. E. (2013). Leadership: A communication perspective. Waveland press.

Hartman, J. L., and McCambridge, J. (2011). Optimizing millennials' communication styles. Business Communication Quarterly, 74(1), 22-44.

Johnson, C. E., and Hackman, M. Z. (2018). Leadership: A communication perspective. Waveland Press.

Zulch, B. (2014). Leadership communication in project management. Procedia-Social and Behavioral Sciences, 119, 172-181.

Chapter 5

Role of Body Language in Job Interviews

Sonali Naik

Research Scholar, Department of Agricultural Extension Education, Odisha University of Agriculture and Technology, Bhubaneswar 751003, Odisha

ABSTRACT

The book chapter "Role of body language in job interviews" delves into the pivotal role of body language in shaping perceptions, building rapport, and ultimately influencing interview outcomes. It provides the detailed information of nonverbal cues, their great impact on interviewers' perceptions and candidates' success are illuminated. The abstract begins with the meaning of body language, importance of body language in the interview context, highlighting the effective body languages and also giving information on positive and negative body language. It includes various nonverbal body languages–from facial expressions and gestures to posture and eye contact and also the importance of each element. By providing theoretical insight and practical guidance, this abstract provides the readers with knowledge and tools that helps them to master body language in career advancement. With the power of nonverbal communication, individuals can unlock doors to opportunities, form meaningful connections, and leave a lasting impression in the competitive landscape of job interviews on the interviewers.

Keywords: *Body language, Nonverbal communication, Posture, Job interviews, Communication skills.*

Introduction

In the dynamic sphere of job interviews, where every gesture and expression are under view, becoming proficient in the art of nonverbal communication can make all the difference between success and disappointment. As individuals try very hard to convey their qualifications, skills, their body language often speaks louder in shaping their ability to understand and influencing interviewer decisions in ways both precise and intense.

Beyond the rehearsed answers, this chapter uncovers the impact of body language in job interviews where a confident handshake, eye contact, or subtle nod can convey competence and professionalism. It is about understanding the nonverbal communication of interviewer like their interpretation and reactions and not just about decoding the signals sent by individuals.

Body language is not just an element of nonverbal communication but it's a powerful tool for building rapport, and leaves a lasting impression. From professionals for career advancement to recent graduate stepping into world of interviews for first time, mastering body language provides opportunities and provide success in the competitive landscape of job interviews.

Body Language

Body language refers to the form of nonverbal communication that involves the use of gestures, facial expressions, postures and other body movements that people use to communicate with others. It conveys a wide range of emotions, attitudes, intentions, and personality traits. It plays a crucial role in interpersonal interactions, as it can significantly impact how others perceive and interpret our messages. A form of communication that is independent of language and culture. Individual can learn body language through practice so that he can act in interview however he wants. It can help in building rapport, convey confidence, navigate social interactions when body language is understood effectively. The study of body language is also known as kinesics. Although body language is an important part of communication, most of it happens without conscious awareness.

Importance

1. It helps in maintaining a professional appearance.

Effective understanding of body language can help in maintaining a professional appearance. Body language is used to create a physical presence that projects honesty, efficiency and productivity. These kinds of appearance are beneficial in career development opportunities also it sets a positive effect on colleagues.

2. Building rapport

The ability of an individual to use their body language positively and build rapport in professional settings, which can help them connect, whilst showing interest and reception. These are beneficial where there are multiple viewpoints.

3. Conveying confidence and competence

More confident body language like direct eye contact, confident posture, *etc.* shown by the individual are perceived as more competent and capable by interviewers.

4. Encouraging team spirit

Team dynamics can be influenced by body language that is important for professionals who work together. It helps the professionals to communicate efficiently and also team spirit is supported because it helps everyone to work simultaneously.

5. Improving leadership productivity

Confidence in the body language displayed by the leader is useful in building trust within a team. Body language is used by the leader while explaining and clarity in speaking. A leader looks more receptive which is an important element to provide genuine support to a team.

6. Influencing decision making

In order to supplement evaluation of candidate's qualifications and responses, interviewers rely on non verbal cues. A candidate's verbal messages can be reinforced by positive body language which leavers a lasting impression and influencing decision making.

Effective Body Language

- ☆ **Eye contact**- It is a sign of confidence. One third of the time eye contact must be maintained during conversation and should be maintained two third of the time while listening.
- ☆ **Handshake** – It is a common gesture in establishing trust and rapport. While meeting someone for the first time, make an eye contact and give a firm handshake.
- ☆ **Body posture** – The individual must sit straight and lean slightly towards the interviewer which would make them appear engaged and interested in conversation.
- ☆ **Listening** – Nodding in conversation will help to show that the individual is taking the conversation seriously.
- ☆ **Hand gestures** – It can be as powerful as words. Overdoing hand movements may distract the interviewers.
- ☆ **Tone of Voice** – Clear communication can be maintained by a gentle tone of voice. Vocal pitch and volume must be taken care of. Monotonous pitch would lead to lack of attention, so individual should aim to keep voice pitch and intensity varied.

Types of Body Language

1. Positive body language
2. Negative body language

Positive Body Language

Positive body language refers to the non- verbal cues which can communicate how people feel. Whether or not you are verbally expressing but your facial expressions how you are holding sends a message to the people. To communicate effectively in work place you must understand your own body language as well as your colleague's too.

Here are six body language techniques that some people consider positive:

1. Eye contact

Active listening and engagement can be indicated by maintaining eye contact. Respect and understanding is also represented, which creates a bond between the communication parties. Considering the cultural differences in your workplace eye contact should be maintained, as some culture finds direct eye contact as disrespectful.

2. Head nod

Nodding head in conversation helps the speaker to demonstrate that they are actively listening to the speaker. It encourages the speaker to respond positively. Enthusiasm is seen with a head nod and a smile and which encourages engagement. Someone can appear approachable and which creates a bond that encourages honest and effective communication.

3. Firm handshake

A firm handshake by an individual indicates their confidence and respect. In a firm handshake, other person's grip must match. It is a common greeting.

4. Open palms

Openness and honesty is implied by open palm. It gives the impression that a person is comfortable and accepting what they are receiving. Interviewee with open palm can signal the interviewer that they are interested in receiving the information openly. Genuine and practical communication can be created by this openness when someone feels comfortable while communicating.

5. Upright and open posture

Interviewee with upright and open posture denotes keeping trunk of body open and crossed arms or legs are avoided. Friendly and willing signal with trustworthy personality is reflected from this body language.

6. Leaning in during speaking and listening

Maintaining distance between people while talking is one of the non-verbal cues. Leaning in and facing the interviewer while giving interview shows that interviewee is interested in the conversation.

Negative Body Language

Negative body language refers to non-verbal cues and gestures that are conveying discomfort, defensiveness, and negativity. These can be tense posture, not maintaining eye contact, crossed arms, *etc.* It indicates discomfort, disagreement, disinterested in a conversation.

Here are five body language techniques that are considered by the people negative:

1. Tense posture

Lack of confidence, disinterested and laziness is conveyed by tense posture. Enthusiasm and engagement of the interviewee is not seen.

Solution- Sitting straight and maintaining a good posture with shoulders back which reflects the interviewee's confidence and attentiveness.

2. Avoiding eye contact

Lack of confidence, nervousness, discomfort is identified if individual avoids eye contact.

Solution- A natural eye contact should be made by the interviewee where there should be balance between looking away occasionally and eye contact.

3. Fidgeting

Twisting hairs, tapping of legs, making small movements are the signals that detects nervousness, anxiety and impatience to the interviewers. It creates a bad and uncomfortable impression.

Solution- Performing deep breathing a relaxing exercises before interviews for the management of nervous system. Try to stay still without making small movements which reflects confidence.

4. Closed posture

Closed posture with crossing of legs or arms signifies resistance and unwilling of openness which creates a barrier for communication between interviewer and interviewee.

Solution- Relaxed arms with hands on the table and keeping the body language open helps in avoiding the barrier and crossed arms with closed posture must be avoided.

5. Absence of facial expressions

An interviewee without any facial expressions shows that he is not interested in the interview being taken which will hinder the judgment of emotions and reactions of interviewee.

Solution- Natural smile with pleasant facial expressions should be maintained throughout the interview showing enthusiasm and interest. Occasional nodding should be done which shows attentiveness and interest.

Conclusion

In wrapping up the exploration on the chapter "Role of body language in job interviews" it is concluded that Success in job interviews can be achieved by an individual when he masters in body language. Body languages, non- verbal communication speaks louder than those of words. Each and every element from facial expressions, eye contact to maintaining postures and gestures play very important role in interview point of view. Maintaining eye contact demonstrate sincerity and attentiveness, open posture signifies friendliness, open palms signifies open and comfortable to accept what they receive, firm handshake indicates confidence and respect, head nod indicates shows interest and enthusiasm of the interviewee, all of these body languages makes an everlasting impression of the interviewee in front of the interviewer and which creates a positive effect in the interview. Body language of the interviewee makes a sense of connection which enhances communication and creates harmonious atmosphere during interview.

REFERENCES

Barkai, J. L. (1990). Nonverbal communication from the other side: speaking body language. *San Diego L. Rev., 27,* 101.

Ekman, P. (1964). Body position, facial expression, and verbal behavior during interviews. *The Journal of Abnormal and Social Psychology, 68*(3), 295.

Ganguly, S. (2017). Understanding Nonverbal Cues: A Key to Success in Interviews. IUP Journal of Soft Skills, 11(2).

Gotcher, M. A. (1990). *Nonverbal communication in the job interview: first impressions and nonverbal immediacy* (Doctoral dissertation).

Karpaha, N., Lakshmanan, L., and Nagarathinam, D. The Importance of Dress Code and Body Language for Interview and Group Discussion. *Sri S. Ramasamy Naidu Memorial College Sattur 626203, Tamilnadu, India,* 61.

Pease, A. (1981). *Body language.* Camel Publishing Company,.

Prickett, T., Gada-Jain, N., and Bernieri, F. J. (2000, May). The importance of first impressions in a job interview. In *annual meeting of the Midwestern Psychological Association, Chicago, IL.*

Sniad, T. (2007). "It's not necessarily the words you say… it's your presentation": Teaching the interactional text of the job interview. *Journal of Pragmatics, 39*(11), 1974-1992.

Sterrett, J. H. (1978). The job interview: Body language and perceptions of potential effectiveness. Journal of applied psychology, 63(3), 388.

Chapter 6

Ensuring Nutritional Security through Family Farming Practices

Sipra Baral and Rosalin Sahoo

MSc Scholar, Food and Nutrition, College of Community Science, Odisha University of Agriculture and Technology, Bhubaneswar 751003, Odisha

ABSTRACT

Food security as an idea created over the last at least 50 years that principally tended to the developing requirement for the creation and admittance to satisfactory food grains to take care of the world's rising populace. Nourishment security, a later turn of events, was a lot more extensive idea since nutritious and safe weight control plans close by sufficient organic and legitimate social conditions guarantees proper development and improvement in youth and advances wellbeing and forestall sickness in adulthood. Achieving nutritional security via circle of relatives farming practices includes selling various, nutrient-wealthy plants, sustainable farming techniques, and community engagement. This will encompass crop rotation, agroforestry, and promoting local food systems to make certain access to clean and nutritious produce. Additionally, supporting small-scale farmers with training, get entry to markets, and sources can decorate their capability to sustainably produce food even as improving food safety and nutrients for their families and communities. In step with modern-day forecasts, international agricultural manufacturing will must growth by 70 per cent by using 2050 to satisfy the meals needs of a international population of over nine billion. The improvement and implementation of agriculture, meals security committed to sustainable improvement and the advertising of states' commitments to the innovative recognition of the right to cook meals in the regard of national meals protection are key steps in the direction of those goals.

Keywords: *Agroforestry, Crop rotation, Meals security, Nutritional security, Sustainable improvement.*

Introduction

India, which is more prominent than billion crowded, is wrestling with developing imbalances, provincial destitution, yearning and unhealthiness. With greater part livelihoods being ranch based, helping circle of family members cultivating is presently not a 'crude' methodology anyway an essential one. Smallholder ranches are family-based mostly,they make a monster commitment to the worldwide food and sustenance security. It may be circle of family members ranches and the countries wherein they work are various in heaps of strategies and the arrangements introduced for them must be customized for this variety. Farmers want a high enough income to maintain their rural livelihoods or to move to urban areas with the hope for a better lifestyle. To prevent this the rate for their produce and offerings is required. Restrained entry to land and different herbal assets, information, training and financing are critically hindering circle of relatives farming improvement globally. Excellent practices of coping mechanisms must be extensively disseminated. If everyone thinks about circle of relatives farming the first query come into thoughts what is own family farm. As described through United States department of Agriculture (USDA) regulations a circle of relatives farm is a farm that produces agricultural commodities for sale in such quantities so one can be identified in the network as a farm and no longer a rural residence:

1. Produces enough earnings (including off-farm employment) to pay family and farm operating charges, pay debts, and keep the belongings
2. Is controlled by means of the operator
3. Has huge quantity of labour supplied by means of the operator and the operator's own family and may additionally use seasonal labor throughout height periods and a affordable amount of complete-time hired.

As per the FAO, family farming is defined as "a family-managed and directed mode of organising forest and crop production, livestock raising, fishery, and aquaculture, which primarily depends on family labour of both men and women." The family and the holding share, integrate, and coevolve various roles in the social, cultural, economic, reproductive, and environmental spheres.

The Malnutrition Problem in India

In spite of the post-freedom, financial and logical headway made by India, the scourge of unhealthiness endures. Indeed, even while pre-change sicknesses like transmittable illnesses and under nourishment endure, there is developing rate of over sustenance (weight) and related non-transferable infections, the twofold weight.

Food security suggests that "all individuals, consistently, ought to have physical, social and monetary admittance to adequate, protected and nutritious food which meets their dietary requirements and food inclinations for a functioning and solid life" (FAO, 2000, reaffirmed authoritatively in 2009).

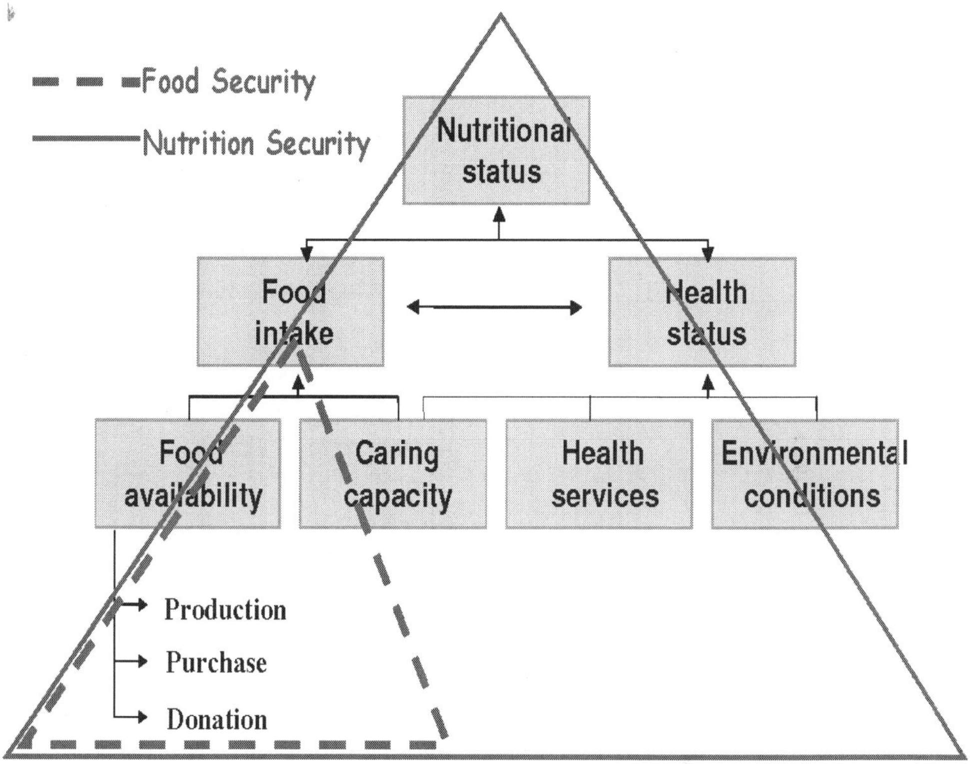

www.semanticscholar.in

Nourishment security goes past food security. For nourishment security, there must be mindfulness and access at reasonable expense to :

1. Security of food and right infant and kid taking care of practices
2. Drinking water safety and infection free climate
3. Outreach for medical services
3. Healthcare outreach.
4. Absorption

Weight loss plan surveys conducted in India reveal that the majority of Indian meals are made entirely of cereal and are, therefore, deficient in micronutrients, vitamins, and minerals (causing hidden hunger). This is because there is insufficient consumption of foods that act as shields, such as vegetables, pulses, and animal products (milk, fish, egg, and meat). In addition to being rich in micronutrients (MN), many B-vitamins, and minerals, pulses are an essential source of protein for vegetarian diets. Millets are a good source of B vitamins and minerals in addition to being high in fibre. Diseases are caused by deficiencies in micronutrients such as zinc, iron, calcium, folic acid, vitamin A, B2, B12, C, and D. In both women and children, iron deficiency anaemia is a prevalent condition.

Role of Family Farming in Nutritional Security

To understand the maximum capacity of Family ranchers in guaranteeing food Security, need to have clear Comprehension of necessities, potential and Limitations of family cultivating and Ideal arrangement climate and specialized help are important. At Public level, there are various

Key factors for an effective Improvement of family cultivating, for example:

Agro-biological circumstances and regional qualities, admittance to business sectors, admittance to land and normal resources,access to innovation and expansion administrations, admittance to back, segment, financial and socio social circumstances, accessibility of specific training among others. Designated rural, natural and social approach mediations on the side of family ranchers are important to roll out substantial improvements and maintainable upgrades.

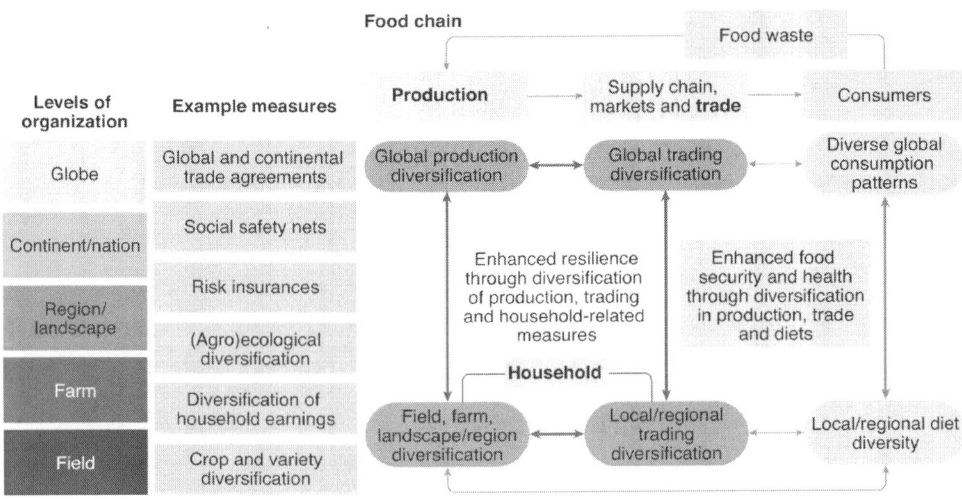

https://www.nature.com/articles/s43016-021-00403-9/figures/1

The Importance of Family Farming

Family farming is the most well-known functional cultivating model in India and in this way critical. The dominant parts of the Indian homesteads are family ranches, passed down starting with one age then onto the next, and add to the financial and natural supportability of country regions. There is a high variety of family ranches in the India, as far as their size, exercises they take part in, accessibility of assets, level of market mix, seriousness, and so forth.

Benefits of Family Farming

It has the following functions for the welfare of all humanity:

1. The Arena is Led by Family Farming

Family farmers are essential to the fight against starvation and malnutrition since they generate 70 per cent of the food produced worldwide. In light of this, small farms consistently outperform large industrial agricultural enterprises in terms of output relative to land area and labour force employed.

2. Farming give rise to Circle of Relatives Properly

Forty per cent of the families in the arena depend on their circle of relatives for their income from farming. Out of the 3 billion people living in rural areas of developed nations, 2.5 billion are members of agricultural households. It is imperative to acknowledge that women account for over 50 per cent of the agricultural labour force in developing nations.

3. Agriculture withstand Poverty

According to the World Financial Institution, a doubled GDP growth in agriculture decreases poverty twice as much as a GDP boom in other sectors. A boom in agriculture, particularly in rural areas, also helps to alleviate poverty in cities by generating more employment and income and supplying healthier and more affordable food.

4. Family Farming Safeguards Biodiversity and the Environment

A wide variety of genetic traits that are well-suited to a range of conditions are used in family farming, including seed and livestock breeds. It further supports the agroecological and traditional methods of farming, which in turn promotes the healthy functioning of ecosystems. Due to their centuries-long experience and understanding of dealing with natural disasters, family farms are thought to be more vulnerable to the effects of climate change.

Family farms generate 70 per cent of the food produced worldwide, making their work essential in the fight against malnutrition and famine. Small farms also produce more per unit of land and with less energy than huge industrial agricultural operations.

5. Family Farming Produces Prosperity

A little less than half of the world's families rely upon family cultivating for their vocation. Of the 3 billion provincial occupants in emerging nations, 2.5 billion have a place with families working in the farm. It is critical to take note of that ladies represent close to half of non-industrial countries' agrarian work assets.

6. Agriculture Battles Neediness

GDP (Gross domestic product) development starting in farming decreases destitution two times however much identical Gross domestic product development in different areas, as per the World Bank. The urban poor also gain from agriculture l and rural development, which creates more wealth and jobs and increases access to affordable food.

7. Family Cultivating Safeguards Climate and the Biodiversity

Family farming is a source of genetic diversity that makes use of seed varieties and animal breeds that have been highly adapted to various environments. It also promotes the use of conventional and agroecological techniques, which helps habitats function well.

Family farming is perceived as stronger to the effects of environmental change due an abundance of information on adapting to catastrophic events, amassed over hundreds of years.

Challenges Faced by Family Farmers

Smallholder and family ranchers are confronted with various difficulties:

1. Change in climate and inconstancy
2. Insufficiency of residency security in a setting of; Increment of contest for water and land(populace development, urbanization) and lacking administration of residential land
3. Minimal admittance to monetary assets, inputs, innovation, preparing, examination and warning administrations, and instruction Cost instability (energy, food, and so on.)
4. Limited admittance to business sectors.

Suggestions

1. Buyer Commitment and Demand Moving

The consumer demand wishes to be moved toward sustainable and maintainable weight control plans. We really want to move to a feasts plate this is better for people and the planet. We can make locally developed millets famous in India similar as imported oats or quinoa by utilizing following the equivalent strategy used by associations. Common society and the wellness organization might associate with online entertainment powerhouses who at any point can shape more sound and maintainable utilization for many thousands.

The overall population framework, through the overall population Dispersion framework, early afternoon food, rail lines catering, city bottles, and public and institutional acquisition, can assist with further developing what something like 70 per cent of Indians are ingesting.

2. Helping Ranchers and Sustainable Farming

To ensure strong wages, we ought to help ranchers' progress nearer to gainful and regenerative horticultural practices. The public endeavor on normal Cultivating is a stage toward this path, but the ordinary financing for supportable farming is under 1 per cent of the horticultural spending plan. Horticulture help should course from enter appropriations to guide cash help to ranchers per hectare of development. It'd advance green utilization of data sources, while empowering a phase playing subject for agroecological practices to flourish. Rural examination and expansion contributions need to distribute a piece in their financial plans to acknowledgment on feasible horticultural practices, that could offer ranchers with the skill and gear they need for manageable cultivating.

3. Constructing Reasonable Farm-to-Fork Value Chains

Making more practical and comprehensive expense chains is crucial for upgrading country wages and ensuring that ranchers get a decent amount of the worth made. Empowering go betweens and organizations you bought immediately from ranchers, explicitly people who notice reasonable and moral practices, is fundamental. Motivators like fair exchange standards can be carried out to sell responsible obtaining. Allowing rancher produce agencies (FPOs) to exchange their produce with various fpos can assist ranchers with getting more prominent charge, because of the reality rancher families in a FPO likewise buy cultivating items.

The Odisha natural Ranchers' connection (OOFA) is an organization of FPOs that produce regular product. The OOFA exchanges normal items with other FPOs in Odisha, as well as with FPOs in different states in India. This has assisted the OOFA with getting a superior expense for its individuals' product, and to arrive at a more extensive client base.

Conclusion

With the expanding stress of populace on water and land resources and prospect of detrimental changes in climate, decentralized circle of relatives farming primarily based on gender, climate and agriculture that is nutrition sensitive agriculture is the way to all.

With respect to the idea of rural land property, India has sufficient solidarity to guarantee family cultivating for sanitation in a more prominent way. The zero-starvation innovative and judicious might be changed over into reality, provided family ranchers are empowered to support their vocations dependent absolutely upon the thoughts of value, nature, financial aspects, and employment. In monetary expressions, family farming is determined to have explicit Pioneering abilities,commercial undertaking proprietorship and control, want and risk lead, strength and character achievement. Family cultivating is routinely extra than a proficient profession, as it mirrors a lifestyle dependent absolutely upon goals and customs about living and paintings. It assures food insurance even while dispensing increasing cultural hopes for feasts safety,first-rate, worth, beginning and variety of dinners. It likewise keeps rustic way of life and adds to socio-monetary and ecological manageability of the provincial areas. The family ranches have an innate adequacy for short assembling development and way to manageable food producing, whenever given the smartest strategy.

REFERENCES

Bamji MS, Murty PVVS, Parimala DS. *Nutritionally sensitive agriculture- an approach to reducing hidden hunger. Eur J Clin Nutr. 2021;75:1001–1009*. doi: 10.1038/s41430-020-00760-60-x.

Fanzo,J.,Covic,N., Dobermann,A., Henson,S., Herrero,M.,Pingali,P., Staal,S. (2020)A research vision for food systems in the 2020s: defying the status quo Glob. Food Sec., 26 (2020), Article 100397, 10.1016/j.gfs.20207.10039

Gill, M.S.; Singh, J.P.; Gangwar, K.S. Integrated farming system and agriculture sustainability. Indian J. Agron. 2009, 54, 128–139.

Gueye,B., Peterson,P. and R. (2009). *Any future needs Connecting worlds of knowledge* Online publication. Pp 1-3.

HLPE. (2017). Nutrition and food systems *A report by the High Level Panel of Experts on Food Security and Nutrition of the Committee on World Food Security.* Rome (2017), p.152 Available at https://www.fao.org/3/i7846e/i7846e.pdf

Swaminathan,M.S.(2014) *Strengthening family farming in India. Financial chronicle.* UN. 2014. International Year of Family Farming

Toulmin, C., and Gueye,B.(2009). Is There a Future for Family Farming. Wiley online library. Pp(1-6)

Chapter 7

Role of ICTs in Family Farming

Likita Das

Research Scholar, Department of Agricultural Extension Education, Odisha University of Agriculture and Technology, Bhubaneswar 751003, Odisha
e-mail: likitadas64@gmail.com

ABSTRACT

In this developing world rural residents still mostly rely on agriculture for their livelihoods. Family farms have many dimensions in terms of generating income, boosting agricultural production, equity and entrepreneurship in both developed and developing nations. In family farming the agricultural land, labor and resources are owned by the same family. Family members typically handle all the facet of farming, from planting to harvesting, managing the farm's finances to caring for cattle. But in recent times farmers and agricultural researchers are still quite apart. Bridging this knowledge gap between farmers and researchers has become a challenge for the extension services. Information and Communication Technologies (ICTs) like televisions, mobile phones, web portals and social media help in retrieving and sharing information to the target audience with in a short duration of time. ICTs can facilitate communication among youngsters and professionals. It also makes agriculture attractive for the younger ones to get engaged.

***Keywords**: ICTs, Web portals, Mobile phones, Social media, Extension services, Family farming.*

Introduction

Technology in agriculture has evolved from animal-drawn ploughs to mobile apps that are guiding farmers in a better way. In developed and rich countries there are resource wealthy farmers and more commercialized markets.

But in under developed countries traditional farming is common with many impoverished farmers following their ancestors' footsteps. For those farmers Extension services are playing great role in bridging the knowledge gap and in achieving the targets of their family farms. Information and Communication Technologies (ICTs) analyze, gather and transmit information like weather forecasting, soil condition, crop health, scheduling irrigation, optimized fertilizer use, pest and disease management *etc.* electronically. According to Heeks (1999) ICTs are the electronic tools that help in recording, processing, storage and reporting of the information. While agricultural extension is bridging the gap between farmer and researchers, ICTs are bridging the gap between extension and farmers. According to agricultural census statistics from ten developing nations, family farms interact with Government extension agents at a maximum rate of 25 per cent. In India it is just little higher than five per cent (FAO, 2014c). This enormous gap between extension and family farms has become a barrier.

ICTs in Indian Agriculture

In India farmers mostly rely on television, radio and community radio, web portals, mobile advisory services for getting the information on time:

Radios and Community Radio

Broadcasting on radio in India began in 1924 by the Madras Presidency Club. All India Radio began as the Indian Broadcasting Company in 1927 in Bombay and Calcutta. Now All India Radio, the public service broadcaster has developed 467 radio stations along with 662 radio transmitters. Chauhan (2007) cites the 1965 'Krishi Prasaran' program was first to disseminate information related to agriculture. Following the success of the 'Intensive Agricultural District Program' in the USA 'information centers were established to spread its benefit to rural areas of India.

Television

In the year 1959, television was launched in India, three decades after its global debut. Doordarshan became the first television channel. The first program for agricultural extension was broadcasted by Doordarshan in collaboration with Ministry of Agriculture, Government of India in 1967. Now a days farmers are getting news on weather forecasting, livestock information, awareness regarding disease and pest management, price fluctuation, consumer demand, supply chain *etc.* Some researchers are also sharing their knowledge on new farming methods on different television channels.

Web Portals

There are many web portals for dissemination of information related to agriculture.

Some web portals for providing general information related to Agriculture are

https://www.ikishan.com/ - initiated by Nagarjuna Fertilizers and Chemicals Ltd.

https://www.ikishan.com/ - Initiated by IIT Mumbai, KVK Baramati (now owned by Agrocom Pvt. Ltd.)

https://www.iffco.nic.in – Developed by IFFCO Ltd. and maintained by NFC

https://www.hul.co.in(i-Shakti) – Started by HUL Ltd.

Some university web portals like https://agritech.tnau.ac.in https://www.pusavarsity.orgn.in https://www.ouat.ac.in

Portals for e-commerce are https://www.agmarknet.nic.in https://agriexchange.apeda.gov.in e-krishivipanan

Mobile Based Advisory Services

Among all the Information and Communication Technologies (ICTs) mobile phones are quite popular because of their easy handling and easy availability. Some telecommunication based advisory services available in India are Kisan Call Centre, mKrishi, Mandi on Mobile, Kisan Help Line, SMS Weather Alert, mKisan *etc.*

Social Media

Social media is the most popular activity on internet. It is quite famous among youngsters for communication. Social media platforms like Facebook, Instagram, YouTube, LinkedIn, BlogSpot *etc.* are quite famous. There are also many demonstration videos available for learning on these platforms. These are helpful for observational learning.

Issues Faced

- **Low literacy rate:** In our country most of the farmers are small and marginal. They are illiterate. They don't know about the uses of these Information and Communication Technologies (ICTs).
- **Lack of awareness among the farmers:** Most of the farmers are unaware of the social media platforms, mobile apps and web portals.
- **Poor connection:** In most of the villages in India there is no proper electricity connection. Poor signal availability is also there which is a major issue for operating of ICTs.
- **Poor management:** Government websites are poorly managed. Databases are not updated regularly. Accessibility is poor.

- **Expensive services:** Information and Communication Technologies (ICTs) are costly. As most of the farmers in India are small and marginal they can't afford internet connectivity, smart phones *etc.*
- **Location specific data:** Most of the data available on the websites, social media platforms aren't location specific. So there is a need of location based data availability to the farmers.
- **No proper availability of assistance:** In telecommunication based advisory services most of the times the queries are unanswered. Farmers don't get any proper solution to their problems.
- **Language barrier:** Most of the ICTs transmit information in English or hindi language. But as most of the farmers are uneducated they always need solution in their local language. So the information available aren't easy to understand for the farmers.
- **Single mobile phone per household:** In the rural areas families are having only one mobile phone. As every member is supposed to get involved family farming a single mobile phone isn't sufficient to get enough information.

Improvements Needed

- **Feedback and Transparency:** Feedbacks should be taken from the farmers regarding the Information and Communication Technologies (ICTs) available, whether they are satisfied with the service or not. This will also help in further development of ICTs. Databases should be updated regularly.
- **Collaboration with institutes:** Collaborating with various research institutes will be helpful in providing proper information to the farmers. This will also ensure timely availability of latest information.
- **Independent of networks:** If the Information and Communication Technologies will be network independent these will be easily accessible to rural areas and areas having poor signal.
- **Regulated Information:** Information shared on the websites and social media platforms should be regulated properly. Available data should be proper. So that farmers won't get any false information.

Conclusion

Family farming is a legacy passed down for many generations. Family farming is not only improving rural livelihoods but also providing solutions to global hunger and unemployment. ICTs are not only enhancing farmers' knowledge but also encouraging youngsters to get involved in family farming. Telecommunication based advisory services have enabled farmers to get quick solution to their queries. Social media platforms like YouTube have facilitated observational learning of the farmers. But these ICTs have limitations also and those areas should be taken into consideration. Providing better connectivity, easy accessibility and creating awareness among the farmers regarding technologies will make ICTs the most efficient extension tool.

REFERENCES

Anwber J. (2012). India among countries with worst internet connections. The Times of India.

Anatoli Marantidou, Anastasios Michailidis and Afroditi Papadaki-Klavdianou. (2011). Information And Communication Technologies as Agricultural Extension tools.

Gerhard Schiefer. (1999). ICT and quality management

Raj Saravanan and Bhattacharjee Suchiradipta. (2015). mExtension- Mobile Phones for Agricultural Advisory Services

Saravanam Raj and Bhattacharjee Suchiradipta. (2015). Role of ICTs in Family Farming: Experiences and Way Forward

Chapter 8

Understanding Communication Patterns in Online Learning Environments

Bibhu Prasad Behera

Research Scholar, Department of Agricultural Extension Education, Odisha University of Agriculture and Technology, Bhubaneswar 751003, Odisha

ABSTRACT

The realm of education has swiftly transformed due to the rise of online learning, highlighting the significance of comprehending communication dynamics within these digital environments. This section delves into the intricacies of communication in online learning, examining both asynchronous and synchronous exchanges, communication between peers and instructors, feedback mechanisms, social engagement, resource sharing, and diverse communication strategies. Establishing effective communication protocols is vital for setting expectations and fostering active participation, with practical advice emphasizing the use of clear, respectful language, active engagement, and leveraging technology for collaborative efforts. The chapter addresses challenges like technical difficulties and inclusivity issues while also highlighting opportunities for innovation, adaptability, and community establishment. By grasping and harnessing these communication patterns, educators can create engaging and inclusive online learning environments that empower students to excel in the digital age.

Keywords: *Adaptability, Communication, Digital, Guidelines, Learning.*

INTRODUCTION

In the rapidly evolving landscape of education, online learning has become increasingly prevalent. With the advent of technology, communication patterns within online learning environments have undergone significant transformations. This chapter delves into the various communication patterns observed in online learning settings, exploring their dynamics, implications, and significance for effective learning.

Dynamics of Online Learning

1. Asynchronous Communication

Asynchronous communication involves interactions that do not occur in real-time. Commonly used in online learning, this form includes tools like emails, discussion boards, and pre-recorded lectures. Unlike synchronous communication, where all parties communicate simultaneously, asynchronous communication provides flexibility as individuals can engage at their own convenience.

One key benefit of asynchronous communication is its adaptability to varying schedules and time zones. Students have the freedom to access course materials and join discussions at suitable times, which promotes accessibility and inclusivity. Moreover, this type of communication encourages well-thought-out responses as individuals have the opportunity to ponder before contributing.

In essence, asynchronous communication plays a crucial role in online education by offering flexibility and necessitating effective strategies to foster engagement and collaboration among participants.

2. Synchronous Communication

In online learning, synchronous communication refers to real-time interactions where participants engage simultaneously, often through live video conferencing, chat rooms, and virtual classrooms. This immediate form of communication allows for instant feedback, collaborative activities, and a shared sense of connection.

One key benefit of synchronous communication is its ability to support lively discussions and interactive learning experiences. Students can ask questions, receive instantaneous answers, and participate in live debates or group projects, promoting active participation and engagement. Moreover, synchronous communication helps combat feelings of isolation by enabling students to connect with instructors and classmates in real time.

Overall, synchronous communication plays a vital role in online learning by facilitating immediate engagement, collaboration, and connection among participants. Effective facilitation and proper planning are key to maximizing the benefits of synchronous communication while mitigating its challenges.

3. Peer-to-Peer Communication

In the realm of online education, peer-to-peer communication refers to the interaction between students on an equal footing, without the intervention of teachers or moderators. This form of communication fosters collaborative learning, enabling students to exchange thoughts, deliberate on ideas, and collaborate on tasks or assignments, thereby enhancing their comprehension of the subject matter. Despite its advantages, peer-to-peer communication in online learning settings can pose obstacles like differing communication methods, uneven levels of involvement, and potential conflicts. To overcome these hurdles, educators can offer advice on efficient communication techniques, set distinct guidelines for teamwork, and facilitate activities that encourage peer feedback and introspection.

4. Instructor-to-Student Communication

In online learning, instructor-to-student communication is the act of instructors reaching out to their students. This communication plays a crucial role in delivering course material, providing feedback, answering queries, and encouraging student engagement.

Effective communication between instructors and students is vital for aiding learning and creating a supportive online educational environment. It entails timely and clear communication of information, including course updates, assignment details, and deadlines. Moreover, instructors must be available to address student questions, explain concepts, and offer assistance throughout the course.

Setting communication guidelines and expectations helps manage student expectations surrounding response times, communication frequency, and appropriate behavior. By establishing these guidelines at the beginning of the course, instructors can reduce misunderstandings and promote productive communication.

5. Feedback Mechanisms

Feedback mechanisms in online learning environments serve as invaluable tools for assessing student progress, identifying areas for improvement, and reinforcing learning outcomes. Through regular assessments, quizzes, and assignments, students receive constructive feedback from both instructors and peers, promoting self-reflection and continuous improvement.

6. Social Interaction

Social interaction plays a crucial role in online learning, contributing to a sense of community, collaboration, and engagement among participants. Discussion forums, social media integration, and virtual communities provide platforms for informal conversations, knowledge sharing, and relationship building, enhancing the overall learning experience.

7. Resource Sharing

The sharing of resources among students and instructors enriches the online learning experience, enabling access to diverse learning materials and perspectives. Documents, videos, links, and presentations are shared to supplement course content, facilitate discussions, and foster a culture of collaborative learning and knowledge exchange.

8. Multimodal Communication

Multimodal communication strategies leverage various communication tools and modalities to accommodate different learning preferences and enhance engagement. Text-based communication, audiovisual presentations, interactive simulations, and multimedia elements are employed to cater to diverse learning styles and preferences, fostering inclusive and immersive learning experiences.

Communication Guideline

Effective communication guidelines play a crucial role in online learning environments by setting the stage for expectations, boosting engagement, and fostering an environment conducive to learning. Teachers must provide clear instructions to ensure that students grasp what is expected of them in terms of participating, submitting assignments, and communicating with their peers. Students, on the other hand, can capitalize on understanding the instructor's availability, expected response times, and preferred modes of communication.

The tempo of communication interactions is greatly influenced by the expectations set for response times. While some courses necessitate immediate replies for seamless real-time cooperation, others allow for more adaptability through asynchronous communication. The extent of participation may fluctuate depending on the format and content of the course; for example, a course that heavily relies on discussions may require more active involvement from students compared to a lecture-centered one.

Effective communication guidelines in education encompass the important use of language. Educators can set standards for employing professional language, adhering to proper grammar rules, and engaging in respectful communication to cultivate a positive learning atmosphere. Additionally, embracing diversity within the student body by accommodating varying language proficiency levels and cultural backgrounds is indispensable in promoting inclusivity and mutual understanding.

Another crucial consideration is that communication styles can differ based on the nature of the course. For instance, students in a self-paced online program may have greater autonomy in deciding how often and in what manner they communicate compared to those in real-time, instructor-led sessions. Furthermore, the subject being taught can impact communication practices; technical subjects often require precise language and in-depth explanations as opposed to liberal arts courses.

Through the establishment of explicit communication guidelines and acknowledgment of varied communication patterns, teachers can foster an environment that is both inclusive and supportive for online learners.

Practical Tips for Effective Online Communication

Effective online communication starts with using straightforward and polite language. Being clear and concise helps convey thoughts effectively in the online realm. Polite communication encourages a positive vibe and encourages productive conversations among participants.

Engaging actively in discussions is crucial for involvement and sharing of knowledge. Encouraging students to share ideas, ask questions, and interact with peers enhances the learning process. Educators can support active engagement by asking stimulating questions, overseeing discussions, and offering prompt feedback.

It's important to ask for clarification when needed to prevent misunderstandings. Students should feel at ease asking questions and seeking further explanations from teachers or classmates. Promoting a culture of open communication where no query is considered insignificant nurtures a supportive learning atmosphere.

In the realm of online education, technology is a crucial factor in fostering communication and cooperation between students and educators. Platforms like discussion boards, video conferencing tools, and messaging applications facilitate instant interaction. Additionally, collaborative document editing tools enable students to collaborate on tasks, projects, and presentations from different locations.

Furthermore, Learning Management Systems serve as centralized hubs for managing learning materials, assignments, and communications. Educators can utilize features like announcements, discussion boards, and chat rooms within the LMS to share important updates and promote engagement.

Beyond just communication, technology also enables collaborative endeavors such as group work, peer assessments, and virtual simulations. These interactive activities not only boost student engagement but also cultivate a culture of teamwork and shared learning experiences.

Utilizing technology efficiently and applying practical communication strategies empowers instructors to establish an engaging online learning atmosphere. This approach encourages lively engagement, nurtures cooperation, and fosters meaningful interactions between both students and instructors.

Challenges and Opportunities in Online Communication

It is essential to tackle the obstacles and advantages of online communication to nurture a supportive learning atmosphere. One obstacle involves overcoming potential hurdles to engagement, like technical glitches, distractions, or feelings of isolation. To combat these challenges, effective strategies include offering technical assistance, establishing a well-organized learning environment, and encouraging active involvement through interactive tasks and discussions.

Furthermore, ensuring inclusivity and catering to diverse learning styles, backgrounds, and capabilities presents another hurdle. Instructors can promote inclusivity by providing various means of presentation, engagement, expression, and participation. This may entail offering alternative content delivery methods, providing accessible materials, and fostering collaborative learning opportunities.

Online communication offers a range of possibilities, enabling the utilization of technology to improve teamwork, imagination, and adaptability. Educators have access to numerous digital platforms and tools that can facilitate collaborative projects, interactive conversations, and multimedia presentations. By prompting students to employ technology innovatively, educators can nurture creativity and enhance digital literacy skills. Additionally, online communication creates avenues for establishing a supportive educational environment. Educators can cultivate student connections by promoting peer engagement, fostering empathy and respect, and providing opportunities for social interaction beyond academic discourse. Cultivating a sense of community and belonging can boost motivation, participation, and overall satisfaction within the realm of online learning.

Conclusion

In conclusion, communication patterns in online learning environments encompass a diverse array of modalities and dynamics, each serving a unique purpose in facilitating effective teaching and learning. By understanding and harnessing these communication patterns, educators can create engaging, interactive, and inclusive learning experiences that empower students to thrive in the digital age.

Chapter 9

Face-to-Face Communication versus Digital Communication

Subhrajit Panda

Research Scholar, Department of Agricultural Extension Education, Odisha University of Agriculture and Technology, Bhubaneswar 751003, Odisha
e-mail: subhrajitpanda78@gmail.com

ABSTRACT

An essential component of all social activities is communication. The usage of computers, smartphones, and social media connectivity increased in the fields of marketing, education, business, and so on following the Covid-19 pandemic. Therefore, the objective of the study is to systematically summaries the past literatures and find out the pros and cons of both the mode of communication. It also highlights the future scope of research in the field of hybrid communication. The findings of the study will be helps to the researchers and the policy makers to take effective decision in their respective field.

Keywords: *Face-to-face communication, Digital communication, COVID-19, Social media, Computer-mediated communication.*

Introduction:

Effective communication is essential to both the survival of humanity as well as the operation of an organization. It entails the production and transfer of ideas, information, opinions, facts, and feelings from one place, person, or group to another (Fatimayin, 2018). Communication has a major role in the guiding function of management. For example, a manager with exceptional qualifications and talents is useless if he cannot effectively communicate. To get the job done right, a manager needs to effectively convey his instructions to his subordinates[1]. Three essential components make up the continual process of communications: the sender, the message, and the receiver (Dayton and Henriksen, 2007). Face-to-face communication (F2F) is when two or more people are seen while speaking.

It's a chance to communicate beyond words while both the person who speaks and listener(s) can observe and adapt based on body language and expression.

Digital communication is the process of exchanging concepts, messages, and data via digital mediums and technologies. Electronic devices and networks are used for both data transmission and receiving. Digital communication enables instantaneous interaction, texting, virtual meetings, and the exchange of digital media over long distances[2].

During the last decades, digital communication increased significantly due to technological advancement (Nguyen *et al.*, 2018). Countries in half of the entire globe were asked or mandated to stay indoors in 2020 as part of an attempt to prevent the COVID-19 virus from spreading. The COVID-19 virus was mainly suppressed by lockdowns and other F2F interaction restrictions, however it also had a negative impact on a number of mental wellness markers, such as PTSD, depression, wrath, and dread.

Governments and experts advised individuals to keep a physical distance from one another while yet keeping social interaction in order to lessen the psychological toll that the limitations were taking. During the lockdowns, a lot of people did in fact use more phone calls, video conferencing, and digital text-based communication to stay in touch with friends and family. However, it's still uncertain if, in terms of people's mental health, these many digital communication options can truly take the place of in-person interactions (Stieger *et al.*, 2023).

The major contributions of this paper are as follow: first, this study summaries the relative past literature. Second, identify the benefits and limitation of the respective mode of communication. And finally, suggest the future scope of the study.

Following sections are organized as follow: section two summaries the past literature while section three and four demonstrate the pros-cons of communication process and major findings of the study.

Review of Past Literature

In this section, the researcher summaries a list of selected past study in the field of comparison between F2F and digital communication. According to Evans et al. (2017) F2F communication adds to the idea of affordances and has distinct social affordances include the ability to converse without the need of digital devices, be physically close to one another, and have conversations on the spur of the moment. Nguyen *et al.* (2018) investigated the effectiveness of F2F interactions over online platforms and found that, F2F communication is more effective though the social media helps to build a strong network. Gruber *et al.* (2022) noted that social media, text messaging, voice conversations, and video calls are examples of digital communication channels that can help individuals connect and communicate, but they cannot replace in-person interactions. According to Stieger *et al.* (2023), the 2020 global lockdowns aimed at containing COVID-19 resulted in a rise in mental health problems. It's unclear if maintaining social contact through digital means actually replaces face-to-face interactions, despite government recommendations to the contrary.

Pros and Cons

Face-to-Face Communication

Face-to-face interaction is one of the most popular forms of interpersonal interactions in the work environment. Despite a number of drawbacks, this communication medium is favored because of its many benefits.

Benefits of F2F communication include:

1. Enabling body language reading
2. Fostering transparency and trust
3. Increasing productivity
4. Facilitates deliberation and judgments and
5. Personal Connection and Motivation

In-person interactions with others provide a feeling of connection and belonging that can boost commitment, morale, and motivation among members of a group or organization (DH Pink, 2011).

Drawbacks of in-person communication:

1. Scheduling a suitable time
2. Meeting arranging expenses
3. Ineffectual in lengthy gatherings

Digital Communication

Connecting and working together has been completely transformed by digital communication. Accepting its potential opens doors, strengthens bonds, and creates opportunities. Use digital media and communication to your advantage if you want to prosper in today's linked world.

At the start of the COVID-19 epidemic, virtual online communication was not novel; yet, this incident sparked a general uptake of computer-mediated communication. While there were undoubtedly advantages to this, such as allowing public life and work to carry on largely as usual and lowering feelings of isolation during lockdowns (Gabbiadini *et al.*, 2020), people soon grew disenchanted with online media and it became evident that they missed F2F communication. Benefits of digital communication include face-to-face interaction from any location in the globe (Castells, 2009), Reduces the need for travel and time, Facilitates communication in groups. Drawbacks of digital communication require the right tools and a reliable internet connection[3]. It includes, 1) Privacy Issues; Since digital communication platforms frequently gather a significant quantity of user data, privacy issues are brought up (Acquisti and Gross, 2006). Individual privacy and autonomy are seriously at stake from the possibility of surveillance and data breaches (Schneier, 2015). 2) Security Risks: According to Dhamija *et al.* (2006), cyberattacks like phishing and malware put the security and integrity of digital communication channels at risk. Malicious actors can abuse consumers due to software flaws and weak encryption mechanisms (Gallagher and Jones, 2014). 3) Misinformation and Fake News; According to Lewandowsky *et al.* (2012), a significant social concern is the quick dissemination of false information via digital communication platforms. According to Pariser (2011), echo chambers and filter bubbles compound the issue by stifling critical thought and reinforcing preexisting opinions.

Conclusion

In conclusion, even though we live in a highly technologically advanced society, especially in developed countries, the benefits of F2F communication for mental health and wellbeing cannot be replaced by the plethora of technical tools and services available. The coming years will reveal if new developments in technology, like virtual reality or 5G, can improve the quality of our social media interactions online to the point where they are on par with in-person interactions. Future study can be extended by conducting some empirical research by collecting the data from respondents.

REFERENCES

Acquisti, A., and Gross, R. (2006, June). Imagined communities: Awareness, information sharing, and privacy on the Facebook. In International workshop on privacy enhancing technologies (pp. 36-58). Berlin, Heidelberg: Springer Berlin Heidelberg.

Castells, M., Fernandez-Ardevol, M., Qiu, J. L., and Sey, A. (2009). Mobile communication and society: A global perspective. Mit Press.

Dayton, E., and Henriksen, K. (2007). Communication failure: basic components, contributing factors, and the call for structure. *The Joint Commission Journal on Quality and Patient Safety*, 33(1), 34-47.

Fatimayin, F. (2018). What is communication. *National Open University of Nigeria.*

Gruber, J., Hargittai, E., and Nguyen, M. H. (2022). The value of face-to-face communication in the digital world: What people miss about in-person interactions when those are limited. *Studies in Communication Sciences*, 1-19.

Koester, A. (2022). Why face-to-face communication matters: A comparison of face-to-face and computer-mediated communication. In COVID-19, Communication and Culture (pp. 115-134). Routledge.

Lewandowsky, S., Ecker, U. K., Seifert, C. M., Schwarz, N., and Cook, J. (2012). Misinformation and its correction: Continued influence and successful debiasing. Psychological science in the public interest, 13(3), 106-131.

Nguyen, M. H., Gruber, J., Marler, W., Hunsaker, A., Fuchs, J., and Hargittai, E. (2022). Staying connected while physically apart: Digital communication when face-to-face interactions are limited. *New Media and Society, 24*(9), 2046-2067.

Schneier, B. (2015). Data and Goliath: The hidden battles to collect your data and control your world. WW Norton and Company.

Stieger, S., Lewetz, D., and Willinger, D. (2023). Face-to-face more important than digital communication for mental health during the pandemic. *Scientific Reports, 13*(1), 8022.

Chapter 10

Role of Communication in Climate Change Awareness

Thadikamalla Dharani

Research Scholar, Department of Agricultural Extension Education, Odisha University of Agricultural and Technology, Bhubaneswar 751003, Odisha
e-mail: thadikamalladharani@gmail.com

ABSTRACT

In the Global South, helping communities adapt and to minimise the effects of climate change requires effective communication. The Global South's climate change communication has been largely concentrated on the amount and caliber of news coverage, but this chapter offers theoretical viewpoints that try to improve the field. Considering the shortcomings of a solitary theoretical framework, a multifaceted approach is suggested, incorporating ideas from social change and health communication among other theories. Structural approach to climate change communication that takes into account scalar interactions and a socio-ecological viewpoint, emphasizing the value of participatory communication in promoting sustainable social change. A global framework that recognizes the interdependences between the global, regional, national, and local. One of the main objectives of public health is to apply science from research to actual change in the world. In order to spur a reaction to climate change, communication is crucial.

Keywords: *Communication, Socio-sustainable, Calibre, Global, Regional.*

Introduction

Widespread and detrimental effects of climate change are seen in plants, animals, humans and the natures environment (IPCC 2013, Zakarya *et al.*, 2015).

Climate change "poses great risks to humankind unless there is a significant global reduction in co2 gas emissions."

According to Zakarya *et al.* (2015)," The Global South is expected to be most severely impacted by climate change; Small island nations and Africa, like Philippines and the Indonesia, are especially thought to be the most exposed and vulnerable to its effects "due to the range of projected impacts, and less adaptive capacity" (IPCC 2007).

The climate change means the variations from ancient years to present year and also we say that changes occurs during this period time is consider as climate change. A shift in the distribution of occurrences around the average (for instance, more frequent but fewer extreme weather events)

Discussions about climate variations are not similar the domain is to select some; they now frequently make print and broadcast media headlines. Governments and the public rely on communication and the media to not only spread knowledge but also to set the agenda for the advancement.

In order to provide the public with greater perspective for the catastrophic hurricanes and droughts that will unavoidably come, there needs to be a more continuous stream of smaller, positive stories about climate change throughout the year. Given the overwhelming dominance of clickbait, algorithms, and manufactured anger in today's media landscape, one could argue that climate change is not urgent enough to warrant public attention.

Use Authoritative Scientific Information

When it comes to climate change, misinformation and disinformation are pervasive and pose significant challenges to solving the problem. Disinformation and false information cloud perceptions of climate science and solutions, causing misunderstandings and frequently resulting in inaction or even negative action. "Misconceptions of the scientific consensus, uncertainty, disregarded risk and urgency, and dissent have been contributed by rhetoric and misinformation on climate change and the purposeful undermining of science," the Intergovernmental Panel on Climate Change (IPCC) states.

Convey the Problem and the Solutions

It's crucial to explain the scope of the climate catastrophe, but doing so can make it seem overwhelming, which makes people uninterested and turn away. A positive message that emphasizes the answers can help people overcome disenchantment and "crisis fatigue" and encourage them to get involved.

Mobilise Action

Every hand must be on deck. It will take everything short of a total revolution in the way we produce, consume, and travel to reduce greenhouse gas emissions to net zero by 2050 and to half that amount by 2030. According to surveys, the majority of individuals in developed economies are willing to make changes in their own lives, and most people worldwide want their governments to act.

Climate Journalists and their Role

The main aim of the media is not always equally successful. Journalism theory views them as the "gatekeepers" of the news flow, having the final say over the matter, persons involved, and situations that is covered by the social network. As a result, researchers have been interested in the role that journalists play in communicating about climate change. During surveys conducted among media pepole and in-depth ethnographic subject conducted in news rooms are mostly not present, the majority of studies is content estimate the capacity of medias outcome and derive information of the journalists is there.

Who "Climate Journalists" Are

The socio demographic traits and opinions of "journalists in climate" are poorly understood, although research has revealed that they are a broad collection of experts with widely differing authorities in between the media. Given that climatic variations is a multidisciplinary issue encompassing scientific. Many number of media desks and their journalists take up the subject because of its political, economic, cultural, and other dimensions.

As a result, few media people who write about climatic variations are specialists in the field. Nonetheless, "climate journalists" appear to have gained some understanding of the subject over time. Though an early US journalist study revealed significant gaps in the journalists' understanding of climate change, subsequent research produced different findings and demonstrated that journalists' awareness of the topic.

How Climate Journalists Work

The nature of climate journalism varies greatly throughout nations in several respects. For instance, it has been demonstrated that the sources they typically use vary significantly depending on the national journalism cultures of the various countries. For example, Experts of the main sources of information about changes of climate in Germany and India, while media and news organizations from other countries, as well as local political elites, appear to be more significant in developing nations. Other aspects of journalists' jobs appear to be fairly consistent across national borders: among other things, journalists' workdays involve identifying newsworthy subjects and emphasizing their newsworthiness in their coverage, all the while abiding by standard journalistic practices like conducting in-depth research and giving readers a fair and impartial account of the subject.

Responsibility of Media and their Role

The universe is not good these days since disasters might happen at any time. Reducing the amount of property damage and human casualties brought on by catastrophes is an important goal that is currently garnering attention globally. It is now widely accepted that the body of knowledge and technology that can be used to mitigate disasters has expanded so much that, with a coordinated, cooperative international effort, it may be possible to save people and lessen human problems, displacement, disaster effect, and economic losses just by improving awareness, and information. Long before a tragedy really occurs, the media should demonstrate its readiness, emergency management, and safeguarding of vital infrastructure. Media have now adapted to needs, behaviours of modern world.

Conclusion

As a result, "the lack of constant attention paid to climate change" [3] and the ineffectiveness of the message hinder the examined medium from accomplishing its goal of evoking a communal reaction and a behavioural shift. Furthermore, information about the health hazards linked to climate change should be presented as a public health concern and should be complemented with expert advice and action items.

REFERENCES

Boykoff, M.T., 2011. Who Speaks for the Climate? Making Sense of Media Reporting on Climate Change. Cambridge University Press, Cambridge.

bregon R (eds) The handbook of development communication and social change. Global handbooks in media and communication research. Chichester, West Sussex, UK; Malden, MA, USA: Wiley-Blackwell. http://alltitles.ebrary.com/Doc?id=10830394

Dutta MJ (2011) Communicating social change: structure, culture, and agency. Routledge, New York

Evans HC (2015) Climate-change awareness and online media in Zimbabwe: opportunities lost? In Halvorsen T, Ibsen H, M'kumbuzi VRP Knowledge for a sustainable world: a Southern African-Nordic contribution, 163

Hemer O, Tufte T (2005) The challenge of the glocal. Glocal Times(1) http://biblioteca.clacso.edu.ar/clacso/coediciones/20100824061923/glocal.pdf

Jan van Cuilenburg, Denis McQuail, "Media Policy Paradigm Shifts, Towards a New Communications Policy Paradigm", European Journal of Communication, Vol. 18, No. 2, 181-207 (2003)

Maibach EW, Nisbet M, Baldwin P, Akerlof K, Diao G. Reframing climate change as a public health issue: an exploratory study of public reactions. BMC Public Health. 2010;10:299.

Mefalopulos P (2005) Communication for sustainable development: applications and challenges. Media and glocal change. Rethinking communication for development, 247–260

Sangeeta Rajesh, Editor, "Climate Change and Media Responsibility" Vidura-Journal of Press Institute of India, volume 2 special issue 1, January- March 2010, pages 1-15

Servaes J (2008) Communication for development and social change. Sage Publications India

Woodward A, Smith KR, Capbell-Lendrum D, Chadee DD, Honda Y, Liu Q, Olwoch J, Revich B, Sauerborn R, Chafe Z, Condalonieri U, Haines A. Climate change and health: on the latest IPCC report. Lancet. 2014;9924:1185–9.

Zakarya GY, Mostefa B, Abbes SM, Seghir GM (2015) Factors affecting CO2 emissions in the BRICS countries: a panel data analysis. Procedia Econ Financ 26:114–125

Chapter 11

Agricultural Extension and Advisory Services for Climate Smart Agriculture (CSA)

Anjana

Research Scholar, Department of Agricultural Extension Education, Odisha University of Agriculture and Technology, Bhubaneswar 751003, Odisha

ABSTRACT

Climate change is probably the biggest environmental concern affecting humanity today in the fight against hunger, malnutrition, disease, and poverty. Agriculture is the most sensitive and vulnerable sector of the economy to the consequences of global climate change since it depends on regional climate elements including temperature, rainfall, and soil health. A more "climate-smart" agriculture is required to reduce the consequences of climate change and increase agricultural revenues and output in a sustainable way. In order to promote cooperative learning among farmers, researchers, and extension agents and to widely distribute CSA methods, climate-related information, technologies, and practices must be customized to local conditions. To determine the proper farming technology and practices needed for CSA, site-specific evaluations are required. Therefore, extension may be extremely helpful in helping farmers deal with the varied effects of global climate change by employing the appropriate method to increase awareness and enlighten farmers about the various adaptation and mitigation strategies.

Keywords: *Extension approaches, Climate, Climate change, Climate smart agriculture, Rural advisory services.*

Introduction

The 21st-century greatest threat to development is climate change. Climate change has wrought severe and possibly irreversible havoc on the biological, ecological, and geological systems of our world. Climate patterns can have an enormous influence on agriculture and food production because of changes in habitats, the spread of pests and plant diseases, the distribution of agro-ecological zones, and the patterns of ocean circulation. Consequently, amid unfavorable climatic conditions, it is essential to increase agricultural production at the same time as reducing yield variability. Implementing "Climate Smart Agriculture" (CSA) techniques more broadly is one recommended tactic to achieve this (FAO, 2010). The three primary tenets of CSA are: mitigating the effects of climate change and enhancing resilience; decreasing and/or eliminating greenhouse gas emissions relative to traditional methods; and boosting agricultural incomes and output in a sustainable way (FAO, 2013a). Rather than focusing on particular farming methods, CSA is distinguished by its desired results. Social science research on the social effects of these CSA practices is scarce, despite a wealth of studies on the biophysical side of climate change. Consequently, in order to influence farmer behaviour or provide them with a range of location-specific adaptation and mitigation strategies, a strong extension network is integral to climate-smart agriculture.

What is Climate Smart Agriculture (CSA)

Since agriculture is a human activity sector that is most vulnerable to climate change it is given significant attention in the global climate change agenda. Agriculture, along with forestry and fisheries, must become "climate smart" in order to mitigate some of the complex issues brought about by climate change. This entails increasing agricultural productivity and incomes in a sustainable manner, ensuring resilience to climate change through adaptation, and minimizing or eliminating greenhouse gas emissions whenever feasible (FAO, 2013a). According to Lipper *et al.* (2014), climate-smart agriculture is a strategy for adapting and refocusing agricultural development in light of the new realities of climate change.

The Three Pillars of CSA

1. **Productivity:** With no detrimental effects on the environment, the goal of CSA is to raise agricultural productivity and incomes from fish, cattle, and crops in a sustainable manner.
2. **Adaptation:** Through increasing farmers' ability to adapt and thrive in the face of shocks and longer-term stresses, CSA seeks to lessen their exposure to short-term risks while simultaneously enhancing their resilience.
3. **Mitigation:** CSA should assist in lowering and/or eliminating greenhouse gas (GHG) emissions wherever and whenever practicable.

This suggests that emissions from agriculture-related deforestation should be minimized, and soils and trees should be managed to optimize their capacity to serve as carbon sinks and absorb CO2 from the atmosphere.

CSA and Rural Advisory Services

Strategies and agricultural methods that support climate smart agriculture are part of the promotion of climate smart agriculture. RAS facilitates the activity changes that will encourage agricultural communities to embrace climate-smart agricultural practices by acting as a link between farmers and new information and tool sources. For several reasons, RAS is ready to assist farmers in promoting climate-smart agriculture in an efficient manner;

- RAS staff members no longer maintain operational ties with rural communities, especially when it comes to the sector level. Only RAS has staff members available around-the-clock to engage directly with rural communities.
- The assets of farm households, the social dynamics within and between communities, and the adaptive abilities of farmers are all included in RAS understanding.
- In order to fully scale up climate-smart agriculture, RAS providers may be required to act as a mediator. To harness the proper technologies, practices, and agricultural products, this calls for new levels of innovative thinking and broader expertise. Therefore, in many cases, it will entail encouraging communication and exchange of knowledge among a wider range of stakeholders carrying out a variety of roles across the policy and practice spectrum.
- RAS have served as a bridge between farmers and research.
- RAS frequently helps rural communities cope with crises during and after natural disasters and extreme weather events by offering relief and getting involved in the reconstruction of rural livelihoods.

By innovatively sharing knowledge on production techniques for climate adaptation and climate-related technologies, Rural Advisory Services (RAS) enables the implementation of climate-smart agriculture (CSA). Here are a few important ways by which extension providers can support CSA:

- **Research and Innovation**: Innovative CSA practices can be developed and promoted by EAS in partnership with research institutions. This includes assessing the efficacy of climate resilient technologies, developing new crop varieties, and carrying out on-farm trials. In order to promote the adoption of modern methods that increase productivity and climate resilience, EAS specialists can disseminate research findings and assist farmers in participating in pilot projects.

- **Climate advisory service access:** Farmers can have easier access to climate services and information with the help of EAS. In order to manage climate risk, this entails setting up weather monitoring stations, offering real-time weather updates, and creating decision-support tools. Additionally, EAS can improve the connections between weather service providers and farmers, giving farmers access to early warning systems and localized climate forecasts.
- **Awareness and Education:** EAS can help farmers become more aware of how climate change affects agriculture and why adaptation is necessary. EAS can assist farmers in realizing the value of CSA practices by offering data on weather forecasting, climate patterns, and climate-related hazards. To teach farmers about CSA methods, such as conservation agriculture, agroforestry, water management, and crop varieties resistant to climate change, EAS experts can conduct training sessions, workshops, and field demonstrations.
- **Climate Vulnerability Assessment and Adaptation Planning:** Farmers can create plans for farm-level adaptation and conduct assessments of climate risk with the help of EAS. Farmers can determine appropriate adaptation strategies and prioritize investments by evaluating how vulnerable their farming systems are to climate change. EAS specialists can offer advice on insurance, risk diversification, and emergency preparation, enabling farmers to make well-informed choices.
- **Technical Support:** Technical advice on CSA techniques specific to regional agro-ecological circumstances can be provided by EAS specialists. In order to increase resilience to climate change, information on suitable crop selection, planting schedules, irrigation techniques, soil conservation measures, and pest management tactics must be provided. Additionally, EAS can provide farmers with advice on how to use climate-smart technologies like precision farming methods, weather monitoring tools, and climate-resilient infrastructure.
- **Agricultural Community Learning Circle:** EAS can help create forums for farmer-to-farmer knowledge sharing where seasoned farmers can discuss their implementation of CSA practices and the lessons they've learned. Through these platforms, farmers can share local knowledge, pick the brains of their peers, and obtain useful insights into effective strategies for climate adaptation. To encourage this knowledge sharing, EAS experts can plan study tours, farmer-led workshops, and field visits.
- **Monetary Subsidies and Rewards:** EAS can help farmers obtain incentives and funding to adopt CSA practices. Information on current government initiatives, financial aid, and climate finance opportunities are to be provided. The development of project proposals and obtaining

funding for clinic-resilient infrastructure, supplies, and training can be assisted by EAS specialists in farming.

- **Enhancing Farmer Group Capacity:** Farmer cooperatives and organizations, especially those run by families, can benefit from the establishment and growth of EAS. Family farmers can access resources, inputs, markets, and training collectively by fostering collective action through EAS. By strengthening their business and negotiating abilities, EAS specialists can help these organizations become more capable so they can demand support that is climate-resilient and participate constructively in value chains.

Extension Approaches used in CSA

- **Climate-Smart Villages (CSVs):** CSVs stand as exemplars of developed rural communities or models of localized initiatives aimed at ensuring food security, facilitating adaptation, and fortifying resilience against climatic pressures. Village development plans, climate-smart technology, local knowledge and institutions, and climate information services are their four main components. The site of a CSV is determined by the climate risk profile of the area and the engagement level of local government and farmers. There isn't a predetermined list of interventions or standard procedure. Rather, the emphasis is on customizing a combination of interventions that complement one another and are appropriate for the particular local situation.
- **Climate Farmers Field Schools (FFS)**: With a focus on experiential learning, the Farmer Field School (FFS) is a non-formal, participatory extension approach that prioritizes farmers' needs (FAO, 2002). It gives farmers a low-risk environment in which to test out novel agricultural management techniques, have conversations about and absorb lessons from their observations, and enhance their ability to make both individual and group decisions by gaining new, useful knowledge and skills. Additionally, FFSs have incorporated components of adaptation to climate change. For instance, the Climate Field Schools in Indonesia focused on methods of adapting to shifting patterns of rainfall, such as documenting and analyzing agricultural rainfall data and locally gathering water.
- **ICT supported network:** In order to address the issue of climate change, farmers are made aware of the availability of various adaptation and mitigation strategies through the use of mobile phones, videos, radios, and other media. In order to achieve food security and spread "Climate-smart agricultural practices," the ICT-based project known as "e-Arik" (e-agriculture) was started in Arunachal Pradesh, India, in 2007.

- **Climate Training:** Extension service providers ought to receive training on climate-smart agriculture because it's a relatively new and developing topic. Extension service providers should be knowledgeable about the terms used in the field and skilled in analyzing and interpreting scientific data. Training can be given on a number of topics, including agro-meteorology.
- **Contingency Crop Planning:** For crop planting, irrigation timing, and plant protection measures, farmers can greatly benefit from this type of calendar. Based on regional weather data, it is generated with the help of the agro-meteorological department. It is interactive in that crop planning is heavily influenced by scientists' scientific knowledge as well as farmers' local expertise of the crops and different farming techniques.
- **Climate Awareness Mass Media Campaigns:** Approaches for mass extension campaigns are highly appealing due to their high reach. For instance, an interactive mass extension program in Malawi has demonstrated the beneficial effects of radio messaging on farmers' comprehension of climate change and its consequences, high-quality compost production techniques, and the significance of compost for resilience building.

Projects on CSA in India

National Innovations on Climate-Resilient Agriculture (NICRA)

The National Innovations on Climate-Resilient Agriculture (NICRA) network project was initiated by the Indian Council for Agricultural Research (ICAR) in 2011 to address the challenges posed by climate change. The undertaking strives to create and advance climate-resilient agricultural technologies, especially in regions of the nation that are susceptible to extreme weather events. The main focuses of this project are capacity building, sponsored/competitive grants, technology demonstration, and strategic research.

Climate Change Knowledge Network in Indian Agriculture (CCKN-IA)

MANAGE, the National Institute of Agricultural Extension Management, is jointly implementing the "Climate Change Knowledge Network in Indian Agriculture" (CCKN-IA) project with technical assistance from German Development Cooperation (GIZ). Creating a strong network exchange system to efficiently share knowledge among various stakeholders in three Indian states–Maharashtra, Jharkhand, and Odisha–is the aim of the CCKN-IA program. This is a pilot project. In order for farmers, extension agents, and policy makers to better adapt to climate change, it is intended that they have improved access to up-to-date, trustworthy, timely, and situation-specific information. These cooperative initiatives help to provide multimodal, two-way exchange processes and creative solutions to efficiently address the needs of farmers in real time.

e-Arik – An ICT based Project

e-Arik (e-agriculture) was an ICT-based project Initiated in 2007 in Arunachal Pradesh, with the dual goals of achieving food security and spreading "Climate-smart agricultural practices." Climate wise farming strategies targeted the project area's crops, particularly the Khasi mandarin orange and rice varieties. It was believed that these methods relied on organic technologies, were low-input, and sustainable. A "Village Knowledge Center" with a computer, internet access, printer, scanner, phone, and TV was established in Yagrung village under the e-Arik initiative. Project facilitators were employed at the center to help farmers obtain ICT-based agricultural information. Additionally, a project site was created that provides baseline data from government departments of agriculture and rural development, extensive information on government programs such as farmers' welfare programs, daily market data, weather forecasts, and information on crop growing and other agricultural methods. By acquiring information directly from the portal, offline CDs, or through facilitator intermediaries, farmers could get ICT-based agricultural data or have remote discussions with other agricultural specialists. The study effectively illustrated how ICTs may be used to promote Climate Smart Agriculture practices, with a focus on organic farming and little external input.

Conclusion

The reason extension services are crucial to CSA is because they facilitate the transfer of knowledge to farmers' fields. What is the best combination of different extension approaches to use in CSA? That depends on a lot of factors, like the target population and its geographic distribution, the complexity of the extension messages, the type and variety of data to be collected from farmers, the available technology, and finally the financial resources for extension. On-the-ground extension implementation must also support advocacy and increase decision makers' awareness of the impending threat of climate change for agriculture in order to make extension more climate change responsive and help address the triple challenge of food security, adaptation, and mitigation.

REFERENCES

FAO. (2002). From farmer field school to community IPM – Ten years of IPM training in Asia. Bangkok, Thailand. Food and Agriculture Organisation of the United Nations (FAO), Rome, Italy.

FAO. (2010) "Climate-Smart" Agriculture: Policies, Practices and Financing for Food Security, Adaptation and Mitigation', Food and Agriculture Organisation of the United Nations, Rome.

FAO. (2013a). Sourcebook on Climate Smart Agriculture, Forestry and Fisheries, Food and Agriculture Organisation of the United Nations (FAO), Rome, Italy. Retrieved from http://www.fao.org/climatechange/37491-0c425f2caa2f5e6f3b9162d39c8507fa3.pdf

GFRAS. (2012). The "New Extensionist": Roles, Strategies, and Capacities to Strengthen Ex-tension and Advisory Services. Global Forum for Rural Advisory Services, 1-3. Retrieved from http://www.gfras.org/en/157-the-newextensionist.html

Jha, S., and Singh, S. (2021). Role of Agriculture Extension for Climate Smart Agriculture.

Lipper, L., Thornton, P., Campbell, B.M. and Torquebiau, E.F. (2014). climate –smart agriculture for food security. Nature Climate Change 4: 1068-1072. Retrieved from http://dx.doi.org/10.1038/nclimate2437

Rupan, R., Saravanan, R., and Suchiradipta, B. (2018). Climate-smart agriculture and advisory services: approaches and implication for future. *National Institute of Agricultural Extension Management* (*MANAGE*).

Sala, S., Rossi, F., and David, S. (2016). Supporting agricultural extension towards climate-smart agriculture: an overview of existing tools. *Supporting agricultural extension towards climate-smart agriculture: an overview of existing tools.*

Saravanan, R. Karthikeyan, S. and Vincent, A. 2018. Extension and Advisory Services for Climate Smart Agriculture. MANAGE Bulletin 3 (2018), National Institute of Agricultural Extension Management (MANAGE), Hyderabad, India.

Shelat, K. N. (2014). Climate-smart agriculture, the way forward–The Indian perspectives. National Council for Climate Change. *Sustainable Development and Public Leadership* (*NCCSD*)*, Ahmedabad and Central Research Institute for Dryland Agriculture* (*CRIDA*)*, Hyderabad, 16.*

Chapter 12

Youth and ICTs for Agricultural Developments

Amrita Abhisikta Pusti

Research Scholar, Department of Agricultural Extension Education, Odisha University of Agriculture and Technology, Bhubaneswar 751003, Odisha

ABSTRACT

The exponential growth of information technology has enabled the modern world to be more informed and aware about their close surrounding as well as the distant ones. It will not be wrong if we say that an informed man makes better decision than an unaware man then it will not be wrong. Due to rapid infrastructure development and modernization people tend to move more towards urban area and take up a stable job rather than sticking to the conventional farming sector. Major portion of youth now a days are showing minimum interest in agriculture due to it's uncertain nature. Indulgence of information technology has brought glory to all sector of economy. So why not in agriculture? India is an agrarian economy and majority of the population are involved in agriculture and allied sector. India has a huge manpower that lies in the age group of 18-35. It can be taken as an opportunity to use the talent of youth to include ICTs in agriculture and increase the economic outcome from agriculture and also empowering the youth as well as attract them to modern agriculture.

***Keywords**: Youth, ICTs, Agricultural Developments.*

Introduction

Indian economy is agrarian type as maximum workforce is involved directly or indirectly in agriculture and allied sector. It also contributes a huge share to the country's GDP. The blooming population of 1.4 billion need to be fed and the country's food security needs to be ensured. In order to ensure food security, we just don't need to focus only on production but we have to improvise in every step starting from land preparation, production, post-harvest management, storage, value addition to market information, policy making. Day by day land is reducing. So, in limited land, changing climate, higher demand, a timely information only can enable the farmer to take better decision and head towards a glorified era of modern agriculture

Importance of ICTs in Agriculture

1. ICTs are cost effective
2. It is easily available everywhere now-a-days. Almost everyone has a mobile phone with them and this is a major tool to access any information around the world.
3. The feedback system becomes strong with the use of ICTs. Farmers get benefitted by the information they get and also their feedbacks can be received which can be considered to further improvise the conditions.
4. The timely availability of vital information like weather data or some beneficial information like market trend or market information can be proved to be extensively helpful to the producer.
5. It also makes the information system more efficient, effective and sustainable. For example: previously an extensionist could reach maximum of 1000 or 1500(approx.) people but due to the indulgence if ICTs the reach has extended multifold.
6. ICTs has made the system even more stronger by assimilating the farmers, extension workers, researchers, bank personnels, policy makers. A strong base is prepared for the agricultural development.
7. ICTs is acting like a bridge between the rural and urban people.
8. More and more talents are coming out from rural areas now a days than earlier due to strong ICT system. In COVID-19 pandemic ICT emerged as the important source of connectivity, education.
9. Even if it is impossible to track and reach a person living in a remote area it might be easy with the help of ICT to reach him.
10. The emergence of geographic information system, geographic positioning system, remote sensing, precision farming the productivity can be increased.

Constraints

1. Lack of awareness is a major constraint with the use of ICT. Some handful persons are extremely efficient in using the ICTs upto it's maximum potential.

2. Illiteracy is another constraint that don't allow ICT to grow full fledgedly.
3. Although majority of population are having access to mobile phone still some people are there who are not able to afford mobile phone or laptops to use. For them the ICT is nothing but dream that is yet to be achieved.
4. Many information are available in specific language only the is not understandable by each and everyone out there.
5. Reluctantness of some community to use ICT.
6. Lack of infrastructure is there as we are still in developing stage.
7. Lack of proper policies to regulate the use of ICT in agricultural system and lack of interest to develop new technology for agriculture.
8. Lack of adequate funding to agriculture in developing countries.
9. Majority farmers in India are small, marginal or landless type, who need specific technology according to their need.

Scope of ICTs in Agriculture

ICTs in agriculture has multiple roles in agriculture system.

1. It can be used by farmers for automation in agriculture that enables them to save time, capital and energy which can be used in other area to get more benefits. In this way the income of farmer is increased efficiently. Hence the socio-economic status of farmer is improved. Automation is also helpful in the areas where labour availability is poor or at seasonal peak period when labour shortage occurs.
2. Unemployment in India is a major problem which can be tackled by attracting rural youth to agriculture. So that employment generation is ensured as well as overall production is also increased as more workforce is involved in agriculture. Also the migration of rural youth to urban areas can be minimized.
3. The skill of youth to innovate new technology can be encouraged. Hence the potential of human resource is best utilized.
4. The bargaining power of farmers is increased by making cooperatives through social media platforms. The co-operatives also carry out many other functions like decision for growing crop, financial assistance among farmers, training programmes *etc.*
5. Women farmers are getting knowledge through ICTs and empowering themselves to live a life of their own choice in their own terms. Hence ICTs play a key role in capacity building.
6. Fetching larger market through e-commerce is being possible only due to information technology. The farmer gets more economic benefit than

local market and consumer also gets a product which is not locally grown.

7. It provides access to information and solution for Environmentally sustainable agriculture or climate smart agriculture. In changing climatic condition the new technologies provide useful information as well as assistance to farmer to overcome the problems.
8. The EWS or Early Warning System enables the farmer to plan for unforeseen climatic condition and hence strengthening the disaster management system. It helps to mitigate abiotic as well as biotic stresses.
9. ICTs enable the small farmers to access financial services and better management of risks and empowering them to access the financial services available for them.

ICTs for Youth

1. ICT enables the farmers to overcome financial constraints by empowering them and making them skilled. Various mobile apps, online trainings and courses aid in this.
2. The rural youth is encouraged to work in a technology assisted environment due to their natural affinity towards technology than conventional farming system.
3. It fosters innovation and creativity among youth and also makes them enable to face the challenges efficiently. In this way the youth also come up with extremely efficient innovations that act as base for further technological development.
4. It uses bottom-up approach for implementing a project so that it becomes more fruitful.
5. Formation of virtual groups can help to know about the agricultural practices, crop management practices, market information of other places.
6. The energy and talent of the youth is channelized in proper direction.
7. Youth can bridge the gap between rural and urban as well as developed and developing condition and hence helping the existing community also.

Conclusion

Food is the basic necessity of human as well as all other organism and in a populous country like India agriculture should be given utmost priority to ensure food security to every citizen. Also, the management of explosively growing population and providing them job is also important for the country's economy which can be fulfilled by agriculture. ICT holds potential for agricultural development and sustainability. ICT can increase the productivity, access markets efficiently and ensure food security. However, many challenges like lack of infrastructure, technological illiteracy need to be addressed. So, government, NGOs and private stakeholders should together to bring better policies and initiatives.

REFERENCES

Agena, Maureen. 2011. Making agriculture 'sexy' and profitable to youth. http://www.eagriculture.org/blog/making-agriculture-sexy-and-profitable-youth-blogs-series-agricultureicts-and-young-professional. Updated on June 21,2011. Accessed on April 22, 2012.

Ahuja, Vivek. 2011. Cyber Extension: A Convergence of ICT and Agricultural Development. Global Media Journal– Indian Edition/ISSN 2249-5835 Winter Issue/December 2011.

Barzilai-Nahon K, Gomez R. and Ambikar R. 2009. Conceptualizing a Contextual Measurement for Digital Divide/s: Using an Integrated Narrative (in) Overcoming Digital Divides: Constructing an Equitable and Competitive Information Society. Ferro E, Dwivendi Y, Ramon G. and Williams M. Idea Group Inc.

Crop life International. 2010. Facts and figures– The status of Global Agriculture. CropLife International aisbl. Belgium.

Express news group. 2012. Attracting youth to farming. The Express News Group. http://expressbuzz.com/search/express-news-service. Updated on April 26, 2012. Accessed on April 26, 2012.

Farming First. 2012. E-Agriculture top 10 Agricultural Interests of http://www.farmingfirst.org/2012/01/e-agriculture-top-10-agricultural-interests-of-2011. Updated on January 20, 2012. Accessed on April 24, 2012. 2011.

Flota Sofia. 2012. Facebook: An innovative solution for farmers in India. http://www.eagriculture.org/news/facebook-innovative-solution-farmers-india. Updated on February 21, 2012. Accessed on April 22, 2012.

Gomez, Ricardo and Camacho, Kemly. 2009. Who uses ICT at Public Access Centers?- Age, education, gender and income differences in users of public access to ICT in 25 developing countries. ICIS Conference, SIG GlobDev Workshop, Phoenix, Arizona, Dec 14, 2009.

Haddad W. 2007. ICTs for Education: A Reference Handbook. http://www.ictinedtoolkit.org/usere/p_page.php?section_number=0 Accessed on October 12, 2008.

International Telecommunication Union. 2009. ICTs and Food Security. ITU-T Technology Watch Report

Mittal, Surabi, Gandhi, Sanjay and Tripathi, Gaurav. 2010. Socio-economic impact of mobile phones in Indian Agriculture. Indian Council for Research on International Economic Relations. Working Paper No. 246.

Mugo, Anthony and Vermeulen, Mireille. 2012. The many possibilities of ICTs in African Agriculture. http://www.agriculturesnetwork.org/magazines/global/youth-and-farming. Accessed on April 24,2012.

NAIP. 2011. ICT in Agriculture. http://www.ekrishinaip.in. Last Updated on Friday, 08 July, 2011. Accessed on 28 January, 2012.

National Youth Policy. 2003. Govt. of India.

Prasad Bhanu. 2009. Ossian Agro's Mobile apps: A farmer friendly. Updated on August 3, 2009. Accessed on May 4, 2012. http://nanoganesh.blogspot.in/2009/08/ossian-agros-mobile-apps-farmer.html.

Chapter 13

Impact of Artificial Intelligence on Interpersonal Communication

Aparna Mishra

Research Scholar, Department of Agricultural Extension Education, Odisha University of Agriculture and Technology Bhubaneswar 751003, Odisha
e-mail: mishraaparna688@gmail.com

ABSTRACT

AI has become a transformative force that is changing many facets of human existence, including communication between people. This essay examines the complex effects of artificial intelligence (AI) on interpersonal interaction, examining both the advantages and disadvantages of this technology. With the efficiency, accessibility, and customisation that AI technologies like chatbots, virtual assistants, and sentiment analysis tools ffer, communication channels have undergone a revolution. But there have also been worries raised about algorithmic bias, privacy, and the decline of real human connection. This paper elucidates the dynamic interplay between artificial intelligence (AI) and interpersonal communication, highlighting consequences for individuals, organizations, and society at large through a thorough analysis of the existing research and empirical evidence.

It also suggests ways to exploit AI's promise while reducing its negative effects on human interaction, as well as directions for future research. In the end, our analysis emphasizes how important it is to promote moral AI practices and protect the fundamentals of real human contact in the digital world.

***Keywords**: Artificial intelligence, Interpersonal communication.*

Introduction

An introduction to artificial intelligence (AI) and interpersonal communication entails comprehending how AI affects social interaction dynamics and communication processes. Interpersonal communication in which an intelligent agent acts on behalf of individuals, influencing language usage, social interactions, and perceptions is known as AI-mediated communication (AI-MC). By creating human-like content–from poetry to cover letters–AI technologies, like ChatGPT, have transformed communication by erasing the distinction between human and machine interactions.

The inability of AI to mimic empathy, emotional intelligence, and the subtleties of human communication, in spite of technological progress, highlights the significance of interpersonal abilities that set people apart from robots. In the era of artificial intelligence, interpersonal communication dynamics are changing, and the value of human abilities like empathy, critical thinking, rapport-building, judgment, cooperation, and collaboration is becoming increasingly apparent. These abilities are emphasized as being critical in an AI-driven future where human traits like empathy distinguish people from robots and promote trust, productive relationships, and leadership. These abilities are necessary for both effective communication and relationship-building.

The need to cultivate and advance interpersonal skills is becoming more and more essential to thriving in a technologically advanced environment as AI continues to automate repetitive chores and procedures. In conclusion, the overview of AI and interpersonal communication highlights how AI technologies are revolutionizing language, social interactions, and the fundamental human abilities needed for successful communication in an artificial intelligence-shaped future.

Communication Technologies' Evolution

The development of communication technologies has had an enormous impact on how people connect and communicate with one another throughout time and distance. The way people communicate has shifted over time due to technological advancements like the iPhone and more recent inventions like smoke signals.

The invention of the telegraph and telephone systems, as well as the most recent advances like the Internet and fiber optics that have reduced the world to a global village where information moves at the speed of light, are among the major turning points in the history of communication.

Technological developments have linked people in ways that were unimaginable in prehistoric times, not *just through enabling* faster and more *effective* communication but also *by facilitating* the *exchange* of ideas, *concepts*, and history across distance and time. A new *phase* in the *development* of communication technology is marked by the integration of artificial intelligence

(AI) into communication operations. Artificial Intelligence (AI) technologies are finding their way into more and *more* communication platforms, *improving* interaction efficiency, personalization, and automation. Beyond simple automation, AI *performs* an *increasingly important part* in communication *through* generative AI, *which creates* content that is human like and affects interpersonal dynamics and language *usage* patterns. AI is having an enormous effect on communication processes as it develops, which begs the question of how these technologies will affect relationships and interactions in the future.

In conclusion, while the advent of AI in communication processes is considered a substantial change towards more intelligent and automated forms of engagement, the historical perspective on communication technology demonstrates the continual innovation that has shaped human connection throughout time.

AI's Effect on Language and Social Interactions

Artificial intelligence (AI) has a substantial effect on language and social interactions; it affects the speed, patterns, and emotional expressions of language as well as how conversation partners are perceived and evaluated. It has been demonstrated that AI technologies, in particular algorithmic answer suggestions like "smart replies," change language dynamics by accelerating conversation and promoting the use of upbeat, emotive language. Conversation partners' assessments of each other's closeness and cooperation may be affected by these shifts in language usage and emotional responses.

However, there are worries about the negative connotations associated with the use of AI, which can lead to more negative evaluations if people are suspected of using algorithmic responses overtly. This is despite the potential benefits of AI in improving interpersonal perceptions and boosting communication efficiency. In a nutshell there are a variety of ways that artificial intelligence (AI) is affecting language and interpersonal relationships. These include beneficial as well as adverse impacts on language patterns, communication dynamics, emotional expressions, and the assessment of conversation partners and viewpoints.

AI's Potential Benefits and Drawbacks for Communication

Artificial intelligence (AI) has both advantages and disadvantages when it comes to communication techniques. Positively, AI makes communication more efficient by speeding up interactions overall, automating duties, and by providing tailored answers. Additionally, AI technologies strengthen communication experiences, leading to favourable user appraisals because AI-mediated communication is efficient and convenient.

On the other hand, the drawbacks of AI in communication centre on fears about job displacement and restricted employment opportunities. As ordinary operations become mechanized due to AI's efficiency and automation, people may grow scared of losing their jobs as a result of the potential effect on employment chances in particular domains.

Furthermore, the need to *rely* on AI for communication processes *begs concerns* about the *dynamics of* the *labor* market in the future and the *necessity* for *people to change* their skill *sets* in *order* to *compete* in a *world that* is technologically advanced. In conclusion, artificial intelligence (AI) in communication may result in positive effects like increased productivity and favorable reviews, but it can also lead to negative effects like job displacement and limited possibilities in the labor market. As such, it's critical to understand and handle the changing role of AI in communication processes.

AI's Social Effects and Public Perception

Artificial intelligence (AI) is perceived differently by the general public for an array of reasons, such as public attitudes toward AI applications, a broad knowledge of AI technologies, and worries about the consequences of AI in many domains like public services, social care, and health. The general public's impressions of artificial intelligence (AI) are influenced by a combination of appreciation for its advantages, unpredictability, potential risks, and worries about topics like data security, replacing human judgment, and decision-making transparency. While AI applications in security, health, science, and education are generally observed favourably for increasing accessibility and efficiency, applications in advanced robotics and online targeted advertising are seen more negatively because of potential ethical and privacy issues.

As media representations and interpersonal interactions can change people's perceptions of *AI technologies and* their impact on society, news coverage and interactions with others play a crucial part in shaping people's conceptions of AI.Public knowledge and comprehension of AI are still insufficient, especially in areas where it has an enormous impact on people's lives, such as eligibility for welfare benefits. This emphasizes the need for greater openness, education, and participation by the public to alleviate worries and foster public confidence in AI technologies. Furthermore, the public's need for autonomous regulation, transparency about the decision-making process of AI systems, and the opportunity to contest AI judgments highlights how crucial it is to guarantee the responsible and ethical implementation of AI in order to conform to public expectations and values.

In summary, public attitudes towards AI are complex, reflecting a balance between perceived benefits and concerns, with the need for policymakers, developers, and society to address transparency, accountability, and ethical considerations to build trust and ensure that AI technologies work for the betterment of individuals and society as a whole

Ethical, Legal, and Societal Considerations

The ethical, legal, and societal considerations surrounding artificial intelligence (AI) highlight the need for increased public debate on AI and the importance of regulating and ethically using AI in communication processes. Public awareness and engagement are crucial in shaping the ethical deployment of AI technologies, ensuring transparency, accountability, and alignment with societal values and norms. Regulation plays a key role in guiding the ethical use of AI, emphasizing the need for clear guidelines, compliance with existing legislation, protection of privacy, and assurance of data integrity in AI systems. Additionally, ethical considerations such as bias, fairness, accuracy, privacy, and legal responsibility are paramount in the development and implementation of AI technologies, particularly in sectors like law where AI can have significant impacts on decision-making and outcome

In summary, the ethical, legal, and societal implications of AI underscore the necessity for public discourse, regulatory frameworks, and ethical guidelines to ensure that AI technologies are used responsibly, transparently, and in alignment with human values and societal expectations

Moral, Legislative, and Cultural Aspects

Artificial intelligence (AI) raises a number of ethical, legal, and sociological issues that call for more public discussion on the subject as well as the regulation and moral use of AI in communication channels. The ethical utilization of AI technology must be shaped by the public's knowledge and involvement in order to guarantee responsibility, transparency, and conformity with social norms and values. The ethical application of AI is greatly aided by regulation, which highlights the necessity of precise standards, adherence to current laws, privacy protection, and data integrity assurance in AI systems. Furthermore, while developing and applying AI technologies, ethical issues including prejudice, justice, accuracy, privacy, and legal accountability are crucial, especially in fields like law where AI might have a big impact.

Research Findings and Case Studies

The research on artificial intelligence's (AI) effects on language creation and interpersonal perceptions, based on the sources given, shed light on important ways that AI affects social relationships and communication dynamics. The aforementioned research underscores the potential detrimental and beneficial effects of algorithmic responses, including smart replies, on linguistic patterns, communication speed, emotional expressions, and interpersonal perception. Although the use of AI in communication processes has the potential to improve interpersonal perceptions, increase communication efficiency, and influence language use, it also raises questions about possible harm that AI could do to social interactions and emotional communication.

AI-generated responses can affect language production, emotional expression, and interpersonal perceptions, altering how people communicate and connect with one other in a variety of circumstances, according to experiments on AI-mediated communication and its implications on relationships. AI in communication opens up possibilities for more efficient, tailored responses, and quicker exchanges, but it also presents issues with communication authenticity, the possible loss of individual communication styles, and long-term implications for language use and interpersonal dynamics. These results highlight how crucial it is to comprehend the intricate interactions between AI systems and human communication processes in order to guarantee that AI is applied sensibly and morally in ways that promote social cohesion and human relationships.

The studies on AI's effects on language production and interpersonal perceptions, in summary, highlight the need for more research, ethical considerations, and awareness of the implications of integrating AI into human communication processes. They also shed light on the complex effects of AI-mediated communication on language use, emotional expression, and social relationships.

Conclusion

To sum up the present study concludes that artificial intelligence (AI) in communication has a substantial impact on language creation, interpersonal perceptions, and relationships. This conclusion is based on the sources that have been presented and the research findings. These studies show that language patterns, emotional expressions, and interpersonal perceptions are among the fundamental aspects of human communication that are impacted by AI-mediated communication. While this has the potential to improve communication efficiency, it also presents new difficulties for language use and emotional expression. Changes in communication dynamics, efficiency, and emotional content can result from the incorporation of AI-generated language, such as smart responses, and can impact how people perceive and interact with each other in different circumstances.

In conclusion, the implications for AI's use in interpersonal communication for the future highlight how crucial it is to comprehend the intricate interactions between AI technologies and people, stressing the need for ethical and responsible AI deployment as well as awareness of how AI affects language use, emotional expression, and social interactions. These results offer useful information for creating artificial intelligence (AI) systems that improve communication in the digital age while maintaining the genuineness and emotional components of human relationships.

REFERENCES

Clark, L. S., and Marchi, R. (Eds.). (2020). Social Media and the Transformation of Interaction in Society. Routledge.

Gillespie, T. (2018). Custodians of the Internet: Platforms, Content Moderation, and the Hidden Decisions That Shape Social Media. Yale University Press.

Hampton, K. N., and Sessions Goulet, L. (2015). Social media and interpersonal communication. In K. Wright and L. Webb (Eds.), Computer-Mediated Communication in Personal Relationships (pp. 101-118). Peter Lang.

Kramer, A. D. I., Guillory, J. E., and Hancock, J. T. (2014). Experimental evidence of massive-scale emotional contagion through social networks. Proceedings of the National Academy of Sciences, 111(24), 8788–8790.

Pang, A. (2019). The Distorted Mirror: Reflections on the Unintended Consequences of Artificial Intelligence. MIT Press.

Ragnedda, M., and Muschert, G. W. (Eds.). (2017). Theorizing Digital Divides. Routledge.

Sunstein, C. R. (2007). Republic.com 2.0. Princeton University Press.

Van Dijck, J., and Poell, T. (2013). Understanding Social Media Logic. Media and Communication, 1(1), 2–14.

Chapter 14

Mediated Communication and Technology

Sushmita Toppo

Research Scholar, Department of Agricultural Extension Education, Odisha University of Agriculture and Technology, Bhubaneswar 751003, Odisha

ABSTRACT

The role of communication is the sharing and receiving of data. Mediated technology has become a new feature to enhance the process of communication flow. Information technology is yet to make its full impact on the agricultural field specially make sure that people have access to these means. Mediated communication is just a new form of communication which includes all the features of the traditional methods of communication. Computer mediated communication makes it possible for individual to share and receive information irrespective of time and space. The growth of the mediated communication actually has an impact in our lives.

Keywords: *Computer mediated technology, Information resource, Network transmission, Non-verbal cues.*

Introduction

The word Communication is derived from the Latin word 'communis' meaning common. According to Leagans (1961), Communication is the process by which two or more people exchange ideas, facts, feelings or impressions in ways that each gains a common understanding of the meaning, intent and use of messages. Mediated communication usually refers to mass media communication. Mediated communication is the form of communication which is based on the usage of ICT.

It is very different from face to face communication as it is very private and it can share a large number of data or information within a small duration of time. Mass media communication is often referred to as computer mediated communication.

Computer Mediated Communication

Computer mediated communication can be defined as all the agricultural information practices that are communicated by using computer as the medium. It consists of various websites, DVD's, CD's, programmes that support the decision, presentations *etc.*

Christopher *et al.* (1995) indicated that Ohio State University Extension service offered computer record keeping course for farmers. Few elements related to computer record was given to the farmers. These elements were:

1. General understanding of a computer record, its usage and importance
2. Establishing accounts
3. Usage of categories and classes
4. Developing tax reports
5. Efficiently using basic computer features like calculator
6. Entering split transactions
7. Using the computer for budgeting
8. Using the quick keys for shortcuts.

With the usage of technology the world is constantly evolving and now we spend a lot of our times online. In the communication platform, talking to each other through mediated platforms like (zoom, Instagram, video conferencing, text messaging *etc.*) is referred as computer mediated technology.

There is a huge difference between face to face communication and computer mediated communication. Face to face communication requires the individuals to be in physical proximity with each. In this type of communication we can see the individual's gestures, body language or any other non verbal cues. Meanwhile in mediated computer communication, there is lack of non verbal cues. Even if the message is reaching faster than F2F method sometimes the message can be misunderstood or misinterpreted due to the lack of non verbal cues.

Components of Computer Mediated Communication

According to Laswell, just like the 5W's of communication, for mediated communication also it can be divided into 5 components: Subject, Host, information resource, channel and relations and effect.

Subject

The subject of CMC is human as they are the network providers, creator and recipient of network. As it is different from other forms of communication, the subject has a much closer connection to CMC. The subject *i.e.* human are a part of a social system therefore it bridges the gap between the two systems.

Host

It is the most important part of the CMC system. In a CMC system a host includes both a hardware and a software.

Information Resource

It includes all the messages that is transmitted by the computer network.

Channel

Channel acts as a path for network transmission that makes network transmission smooth. It is the foundation of CMC.

Relations and Effect

After the network transmission, there are some social impact that has been produced. It can be a positive or a negative effect. The purpose of CMC should be focusing on the positive aspect than the passive one.

Characteristics of Mediated Communication

1. There is a chance to get the message delayed due to some asynchronous events of a medium.
2. Lack of Non verbal cues as there is no face to face communication.
3. Without any visual or audio sometimes it can be possible for an individual to get the wrong idea or misinterpret a certain information.
4. It is permanent in nature. We can record the messages, look back at our history *etc.*
5. We can maintain our anonymity in certain websites hence our privacy is maintained.
6. Computer mediated communication enables us to reach to people globally with just a click
7. CMC is a multitasking job. We can talk to many people at once or do some other job as well simultaneously.

Advantages of Computer Mediated Communication

1. It is easily accessible irrespective of the time and place
2. Individuals can use it as per their convenience.
3. It is cost effective. One need not travel across the globe just to deliver a message to someone or vice versa

4. Enhanced features like video calls, text messages, emails make the process less complex and fun
5. It saves a lot of time as compared to the traditional methods of communication
6. It is flexible to use. An individual can use any mode of communication as per their convenience, preference or technical skills.

Disadvantages of Computer Mediated Communication

1. The most important of all, as there is lack of non verbal cues, it can lead to misinterpretation or misunderstandings.
2. If too dependent on this medium, it can act as a source of distraction and hinder our daily activities.
3. Too much dependency can lead to lack of social skills, anxiety, loneliness, *etc.*
4. Even though various websites promise to protect the user data but sometimes due to any hacking or security breaches our data can be exposed and can be misused.
5. Overuse of mediated communication leads to social isolation
6. The mediated communication has the danger of technology glitches, software clashes that may come in the way of communication flow.

REFERENCES

Adhiguru P and Birthal, PS. (2006), ICT in agricultural development, issues and Strategies in *International conference in social science perspectives in agricultural research and development*. February 15-18 2006, IARI, New Delhi, India, 221p.

Christopher, D. Penrose, Janet and Hoilingworth, M. (1995), Collaboration hold a successful record keeping course for small business *J Ext 35:5-9*

https://krishikosh.egranth.ac.in/items/f2c8ad68-0afb-4b96-aec7-34eb22f0e198

Kasdorf, W.E. (2007). The Columbia Guide to Digital Publishing. Suzhou: Suzhou University Press

McQuail, D. (2005). Mcquail's Mass Communication Theory. 5th ed. London: SAGE Publications.

Schramm, W. and Porter, W.E. (2007). Men, Women, Messages, and Media: Understanding Human Communication (2nd edition). Beijing: Peking University Press.

Stanley, J.B. (2008). Introduction to Mass Communication: Media Literacy and Culture (4th edition). Beijing: China Renmin University Press.

Chapter 15

Impact of Digital Platforms on Journalism

Kornu Jhansee

Research Scholar, Department of Agricultural Extension Education, Odisha University of Agriculture and Technology, Bhubaneswar 751003, Odisha
e-mail: jhanseekornu@gmail.com

ABSTRACT

Digital technologies have changed the journalism for better and in negative ways The data age has made many difficulties for each profession. On account of journalism the presentation of data innovation has modified impressively different parts of the profession. The innovative transformation has fundamentally changed the way the public acquires its news and data, and has denied the broad communications of its conventional restraining infrastructure. This paper will investigate convergences between new media, reporting and innovation to upgrade how we might interpret the impact of data and correspondence innovation, extraordinarily web innovation on conventional news coverage. One of the most significant impact of digital forms in journalist is ability to reach to the global spectators. Digital media enables journalists to make more extravagant stories with recordings, infographics, and intelligent content.

***Keywords**: Journalism, Digital Technologies.*

Introduction

Digital platforms has great impact on journalism. With the increase in digital age, the traditional ways to disperse the data and news have been disrupted, preparing for an additional democratized and decentralized media landscape. The field of journalism envelops a few activities for example, news revealing, composing, altering, photography, and broadcasting. Digital media associations generally engage in the act of altering and editing their stories before to their distribution, sticking to the particular criteria of accuracy, quality by each entity. Media practises around the world have been progressively changing because of the current perspective of information and communication technologies (digital technologies) as strategic endeavours, aiming at successfully managing assets to promote organisational, individual,growth, national productivity and development. Obayi *et al.* (2018) guarantee that the introduction of information and communication technology (ICT) has significantly changed numerous features of society, particularly the techniques used for gathering and spreading information. This claim implies that there are numerous ways in which ICT has had an affect on the journalism industry. In order to support this claim, John (2009), Croteau and Hoyness (2003), as cited in Obajuluwa *et al.* (2019), argued that the emergence of new media technology or convergent media has caused a transformative shift in the communication landscape of society, thereby sparking a worldwide revolution within the media industry. Wilson (2014) claims that journalists have adjusted the utilization of advanced innovations in their expert practices, very much like different clients. The quick idea of web news distribution has changed public discussion elements on a social level. Online entertainment powered every minute of every day consistent pattern of media reporting requires quick re activities and continuous updates. Due to this social change, journalists are feeling burden to give news quickly in a computerized world where individuals' capacities to focus are as often as possible short.One of the major impacts of digital platform on journalism is the capacity to reach the global audience. With the increase in digital platforms, different organizations can now contact individuals everywhere, paying little mind to geographic area. This has prompted an additional different and comprehensive media landscape, as individuals from varying backgrounds can now get to news and data from various sources.

What is Journalism?

The act of providing news by gathering information and using storytelling techniques. Journalism is and has been estimated, explored, contemplated and criticized overall by individuals coming from a wide assortment of disciplines. For sure, research about journalism and among journalism has been laid out as a broadly recognized field, especially in the last part of the twentieth century. Worldwide one can track down colleges, schools and universities with devoted divisions, exploration and showing programs in news coverage. The field even has its own global and public journals. This recommends reporting as a discipline and an object of study depends on a consensual collection of knowledge, a broadly mutual perspective of key speculations and techniques, and a between public act of educating, learning and investigating news coverage. Journalism as an academic discipline is still a lot of under basic discussion.

Over the course of journalism, the field has needed to adjust among university and industry, each with its own systematized assumptions and expectations, driving observers to finish up journalism education has wound up as difficult to define; it feels itself disliked by the industry and endured, scarcely by the institute.

New Era of Journalism

Information and communication technology are rapidly changing the act of journalism, similarly as they are changing the way that administrations are conveyed in different sectors (for example, health, political and educational sectors among others). The web is the verifiable medium with the speediest development. The principal innovation of the internet is innovation and digitalization phone and PC advances has fundamentally changed how news and data are delivered and scattered. Globalization, which is right now a popular term, has upset the world and carried with it competition. Individuals' life has changed, either to improve things or for the more terrible, contingent upon which side of the gap they are on. As indicated by Talabi (2011), journalists all over the planet have embraced the web rapidly. It is difficult to exaggerate the commitment of new innovation to further developing journalism as a business substance or as a person. Thusly, there has been an increase in the dissemination of knowledge and workshop training relating to computerized advancements and applications like Wiki, Podcasting, Video-publishing content to a blog, URL, and Web 2.0 (Mawutodki, 2009), as seen by Obajuluwa *et al.* (2019)

Wiki

A kind of site that empowers clients to add, re-move, modify, put, or alter content by visiting a site (news, diversion, and so forth.). Wikipedia, at times referred to as web journalism or electronic journalism, is broadly viewed as the most proficient data and correspondence technology (ICT) stage for taking part in aggregate and cooperative composition inside the domain of online news coverage. It is of most extreme significance to remember that Wikipedia considers momentary altering.

Podcasting

A radio-style broadcast known as a digital recording or pod cast can be downloaded from the web and can be heard on a PC, a mp3, or a Compact disc that has been copied to it. Thus, a digital recording can be utilized for correspondence, occasion data, and radio stations, in addition to other things. Narrating, headings, remarks, sportscasts, audio visits, and virtual visits are further purposes. The most common way of making a digital broadcast includes four stages. They consist of editing, promotion, hosting, recording.

Digital Platforms

Digital platforms can interface individuals, resources and associations determined to work with the center connections among consumers and business as well as guaranteeing a more prominent productivity for the business management. New business ideas, as imaginative new companies, are subsequently made in view of versatility, development and the connections inside the local area around them. The motivation behind this work is to comprehend the development of plans of action brought by unique and creative organizations working through online platforms.

Digital platforms have turned into a significant mode for sorting out a great many human exercises, including monetary, social, and political collaborations. Specifically, the ascent of advanced stages has changed the scene of different ventures like transportation (*e.g.*, Uber, Ola), Programming improvement (eg., Apple, iOS, google android) hospitality (*e.g.*, Airbnb, Couch Surfing).

Major Impact

- ✰ **Accessibility:** Digital platforms have democratized admittance to data, permitting pre-journalism to effortlessly get data, source and research materials online.
- ✰ **Audience Engagement**: Social media and online discussions furnish pre-journalists with platforms to connect directly with their audience, gather feedback, and incorporate user-generated content into their reporting.
- ✰ **Multimedia Storytelling**: Digital platforms enable pre-journalists to use multimedia elements such as podcasts, videos, interactive graphics and to enhance storytelling and convey information more successfully.
- ✰ **Worldwide Reach:** With the web, pre journalists can contact audience around the world, separating geological hindrances and working with diverse trade of information and ideas.
- ✰ **Data Journalism**: Digital platforms provide access to immense amounts of data, engaging pre-journalists to conduct in-depth analysis and produce information driven stories that reveal patterns, trends and insights.
- ✰ **Difficulties of misinformation**: While digital platforms provide opportunities, they additionally act difficulties such like the spread of disinformation and misinformation, requiring pre-journalists to be watchful in checking sources and facts
- ✰ **Citizen Journalism**: Digital platforms have empowered ordinary residents to become journalists by permitting them to report news through blogs, social media, and other online channels. This has extended the sources of news and increased the diversity.

☆ **Interactive content**: Pre-journalists can make interactive content and engaging through multimedia components like recordings, digital broadcasts, infographics, and intelligent guides, improving the narrating experience for audience.

Conclusion

The utilization of computerized advancements in journalism industry has set out open doors and difficulties while likewise advancing the field. It is unarguable that practically all expert works on, including journalism practice, has been straightforwardly or by indirectly impacted by the development of advanced innovations. The result reasons that computerized technology has worked on the expert course of production, news gathering and dissemination in this way, has significantly impacted journalism practice.

REFERENCES

ACCC https://www.accc.gov.au › filesPDF The Impact of Digital Platforms on News and Journalistic Content

Cohen, N.S. (2018), 'At Work in the Digital Newsroom', Digital Journalism, https://doi.org/10.1080/21670811.2017.1419821

Croteau, D. and Hoynes, W. (2003). Media society, indus-tries, images and audiences (3rd ed.). Thousand Vaks: Pine Forge Press

Fedler, F., A. Carey and T. Counts (1998) 'Journalism's Status in Academia: A Candidate for Elimination?', Journalism Educator 53(1): 4–13

John, V. (2009). The media of mass communication (9th ed). Boston: Person Education Inc

Kane, G. C., Alavi, M., Labianca, G., and Borgatti, S. P. 2014. "What's Different About Social Media Networks? A Framework and Research Agenda," MIS Quarterly (38:1), pp. 275-304

Obajuluwa, T. M., Talabi, F.O. and Oluwasola, O. (2019). Knowledge level of new media application tools and op-erational use challenges (A survey of radio and TV jour-nalists in government-owned media organizations). Asian Journal of Advanced Research and Reports, 7(1), 1-7.

Obayi, P.M., Onyebuchi, C.A. and Uwanuakwa, P.C. (2018) Journalists' perception of the influence of ICT in en-hancing journalism practice in Owerri, Imo State. ESUT Journal of Media Studies, 12 (1), 122-128

Raudsepp, E. (1989) 'Reinventing Journalism Education', Canadian Journal of Communication 14(2):1-14

Talabi, F. O. (2011). Technological determinism and its im-plications for Africa media consumers. Agogo: Journal of Humanities, 2, 1-18

Tan, B., Pan, S. L., Lu, X., and Huang, L. 2015. "The Role of IS Capabilities in the Development of Multi-Sided Platforms: The Digital Ecosystem Strategy of Alibaba.com," Journal of the Association for Information Systems (16:4), pp. 248–280.

Chapter 16

Communication Strategies in Health Awareness Programs

Aniket Das

Research Scholar, Department of Agricultural Extension Education, Odisha University of Agriculture and Technology, Bhubaneswar 751003, Odisha

ABSTRACT

Health is a factor which holds primary importance as far as an individual's life is concerned. It has been observed that the number of lifestyle disorders can be prevented largely through proper health awareness programs. While the success or failure of health awareness programs depends on a lot of factors, effective communication is the key to solving the problem of limited success in this field. In determining the success of a health awareness program, it is to be kept in mind that all the individuals, irrespective of community, group or creed should be covered in these programs. The recent studies suggest that communicable diseases like AIDS and non-communicable diseases like cancer could be prevented through early detection and diagnosis. This diagnosis is possible by spreading awareness regarding the symptoms, prevention measures etc. To implement the health awareness programs, an effective communication strategy must be implemented. The strategy must be respectful to the culture and tradition of the community under consideration. The strategy must be simple, concise and must use interactive in nature. Using effective communication strategies in health awareness programs in crucial in preventing fatal and life-threatening diseases.

Keywords: *Health awareness programs, Effective communication strategies, Interactive, Life-threatening diseases, Community.*

Introduction

Health is now the most important aspect of any individual. While the number of lifestyle disorders has seen a meteoric rise in the recent years, people, especially youth have taken the aspect of health education quite seriously. It is important to understand that health education and health awareness is not available in a pill which can be consumed overnight. It can be transferred to the remote areas only through successful health awareness programs. Sincerity, dedication, and effective implementation of these awareness programs is a pre-requite for successful completion of the project. Not a single individual from any community, any ethnic group, any regional tribe, or any metro city should be left out from these programs. Sound health and a fit life are basic rights which all the citizens should enjoy. And health awareness programs can be effective in ensuring that.

The most crucial factor determining the success of health awareness programs is communication. While at times, it becomes tough and poses a challenge to extension workers to reach out to the people in rural areas; having a proper communication strategy beforehand can speed up the process and it also helps in removal of bottlenecks in the application and implementation of health awareness programs.

Importance of Health Awareness Program

A health awareness program is an important tool to ensure sustainable development and improvement in the field of health. Proper research in the field of health is required before developing the required tools and systems for the implementation of the program. According to the WHO, four areas are crucial in a health awareness program, prioritizing health research so that the health needs of the countries can be met, health research systems strengthening, meeting of ethical standards, translating the knowledge into application.

It has been observed that around 48 per cent deaths in India are occurring from communicable diseases and around 90 per cent of them can be prevented through proper and timely health awareness programs. It is unfortunate to note that around 1 million new cases of cancer are reported from this country every year, with most of them being in stage 3 or stage 4 and are deemed uncurable. If only proper health awareness programs could have reached these people in time, the disease could have been treated in time or at least could have been detected in time, in stage 1 or stage 2. While cancer is a non-communicable disease, the case of communicable diseases does not paint a different picture at all. The most serious communicable disease, AIDS, permanently forbids a person from exchanging any kind of fluid or fluid transmitting tool with another person. The various treatment methods for AIDS are still under trial and those available, do not offer a sure-shot solution. How can we forget about the pandemic took the entire world in its tentacles and shook it violently. According to WHO, at least

3 million people lost their lives to this deadly virus. Proper health awareness programs could have easily helped in largely preventing the disease and reducing the number of deaths till the vaccine was made available to all.

Communication Strategies in Health Awareness Programs

Communication is not merely the exchange of emotions or feelings through words. An effective communication strategy must be helpful in sharing knowledge, teaching skills, and encouraging our clients to adopt the prescribed method. Communication is an integral part in almost all the fields wherever transfer of knowledge is required. But health is an area where communication holds paramount importance simply because health is the most important aspect of human life as discussed above. And the sub section of health where communication is used in the most dynamic way, is a health awareness program. The communication strategies used is a health awareness programs must be efficient, effective, to-the-point, easy to adopt and must enable the extension worker to connect deeply with the target audience. The following communication strategies can be adopted for an effective health awareness program:

1. **Proper understanding of target audience** – Health awareness program is not limited to villages. It can be extended to cities, metros, tribal areas, semi-urban areas, outskirts, *etc.* Hence, it is important for an extension worker to understand his target audience first. It enables him to get acquainted with the culture and traditional of the local area.
2. **Using simple language** – The means of communication should be simple, crisp, clear, and concise. Using too many technical words, complex terms, *etc.*, tends to confuse the audience and the health awareness program will be deemed as a total failure.
3. **Using multiple communication methods** – Every individual has different tastes and preferences. While the older people might prefer traditional methods such as community meetings, lectures, *etc.*, the newer generation is more inclined towards digital media like videos, ppts, *etc.*
4. **Interactive communication** – Just one-way lecture is often boring and ineffective. The meetings or lectures should be made interactive. Everyone should be given a chance to interact with the speaker and present his/her own views on the subject matter. Everyone's opinion must be respected and people should be encouraged to participate in the discussion.
5. **Demonstrative activity** – Practical is always more interesting than theory. This holds true in case of communication also. Demosntrative activities such as videos, role-plays, puppet shows, *etc.* are effective ways to catch the attention of the audience and keep them intrigued.

6. **Respecting the culture** – While performing any activity, the extension workers must keep in mind to respect the culture and traditions of that area. Some minor aspects such as conducting the programs during daytime instead of evening, avoiding places like near a temple or mosque, seeking the permission of village elders, *etc.* must be kept in mind before conducting a health awareness program.
7. **Mechanism for feedback** – A proper feedback mechanism should be incorporated in the program for efficient review of the awareness program. Without a proper feedback mechanism, any program is worthless since we do not have any evidence to measure the success of the program. Feedback forms can be distributed to different sections of the audience, such as older and younger people, men, women, and children, *etc.*
8. **Collaborations with health professionals** – The extension agents must seek an opportunity for expert collaboration such as joining hands with doctors or medical officers. This shall build a sense of trust with the audience and they will take the program more seriously.
9. **Highlighting of important steps** – The entire program may not be very important to the audience and hence, important steps must be highlighted. Steps such as changing of syringe before giving an injection, changing of blade before shaving, washing hands before eating, building a toilet inside the house, *etc.* are crucial sanitary measures and hence must be highlighted and repeated so that people will take them seriously.
10. **Evaluation and follow-up** – Consider the feedback of the audience, take up positive criticism and keep updating the awareness program according to the need and current situation.

Conclusion

With change in generation, with change in era, with change in century, one thing must not change, *i.e.* emphasis on health education and awareness. The methods to spread awareness as well as the strategies will keep evolving with time, but the fact can not be ignored that communication plays a key role in spreading awareness regarding health. Effective communication strategies must be formulated and adopted to ensure proper completion of the program. The old saying, 'Prevention is better than cure', holds true even now. And prevention is only possible through efficient awareness, which in turn, is possible through effective communication.

REFERENCES

Colby SE, Johnson AL, Eickhoff A, Johnson L. Promoting Community Health Resources: Preferred Communication Strategies. Health Promotion Practice. 2011;12(2):271-279. doi:10.1177/1524839909333055

Don Nutbeam, Health literacy as a public health goal: a challenge for contemporary health education and communication strategies into the 21st century, Health Promotion International, Volume 15, Issue 3, September 2000, Pages 259–267, https://doi.org/10.1093/heapro/15.3.259

Dorji, Thinley and Wangmo, Shacha and Gurung, Mongal and Tenzin, Karma and Penjor, Tashi and Lucero-Prisno, Don. (2020). Knowledge, attitude and practice towards health research and its utilization among health professionals in Bhutan: Results of a nationwide cross-sectional survey in 2019. Journal of Health Services Research and Policy. 26. 135581962095448. 10.1177/1355819620954481.

Ebina R, Kawasaki F, Taniguchi I, Togari T, Yamazaki Y, Sparks M. The effectiveness of health communication strategies in health education in Kushima, Japan. Global Health Promotion. 2010;17(1):05-15. doi:10.1177/1757975909356628

Health literacy awareness training for healthcare workers: Improving knowledge an intentions to use clear communication techniques, Patient Education and Counseling, Volume 85, Issue 3, 2011, Pages e225-e228, ISSN 0738-3991, https://doi.org/10.1016/j.pec.2011.02.022.

Leslie B. Snyder, Health Communication Campaigns and Their Impact on Behavior, Journal of Nutrition Education and Behavior, Volume 39, Issue 2, Supplement, 2007, Pages S32-S40, ISSN 1499-4046, https://doi.org/10.1016/j.jneb.2006.09.004.

Michael Mackert, Jennifer Ball, Nichole Lopez, Health literacy awareness training for healthcare workers: Improving knowledge and intentions to use clear communication techniques, Patient Education and Counseling, Volume 85, Issue 3, 2011, Pages e225-e228, ISSN 0738-3991

NIEDERDEPPE, J., BU, Q.L., BORAH, P., KINDIG, D.A. and ROBERT, S.A. (2008), Message Design Strategies to Raise Public Awareness of Social Determinants of Health and Population Health Disparities. The Milbank Quarterly, 86: 481-513. https://doi.org/10.1111/j.1468-0009.2008.00530.x

Plimpton S, Root J. Materials and strategies that work in low literacy health communication. Public Health Rep. 1994 Jan-Feb;109(1):86-92. PMID: 8303020; PMCID: PMC1402246.

Ramirez, P., Ramondt, M., and Vougart, P. (2019). Challenges and Opportunities for communication Strategies to improve Environmental Health Literacy. *J Health Commun, 24(1)*, 75–83. https://doi.org/10.1080/10810730.2019.1574320

Chapter 17

Power of Non-Verbal Communication in Public Speaking

Pooja Samal

Research Scholar, Department of Agricultural Extension Education, Odisha University of Agriculture and Technology, Bhubaneswar 751003, Odisha

ABSTRACT

Communication is known as a process of sharing thoughts, ideas, feelings and emotions and understanding the same from one person to another. From ancient cave paintings to present day social media interactions, ways of communication have evolved with time but its essence of connecting, sharing information and understanding one another has always remained the same. There are various ways of expressing oneself, encompassing verbal and non-verbal cues. Verbal communication involves sharing information through spoken or written words where as non-verbal communication involves interacting using wordless messages by means of facial expressions, gestures, body language and other cues that complements or substitutes verbal communication. They are intricately entangled to enhance the clarity and quality of the message. Non verbal cues always act as shadows to the verbal communication adding more depth and meaning to it. Public speaking is more than a skill, an art that not only evolves around transcending words but also gives importance to captivating, inspiring and persuading its audience. This is not possible without the helps of non-verbal cues. Facial expressions, gestures, tone, postures play a pivotal role in engaging audience, enhancing message clarity, building connection and enhancing persuasion and influence. Non-verbal communication acts like a tool for the speaker to create a worthy first impression, convey their intentions with clarity, showcase authenticity and confidence, captivate audience's attention and communicate across cultures. Ultimately, it shows the great potential of these skills to elevate the quality of public speaking and resonance of the audience with the speaker.

Keywords: *Non-verbal communication, Public speaking, Body language, Gesture, Facial expression.*

Introduction

We have always heard, "Action speaks louder than word." The same principle applies to public speaking as well. In the realm of public speaking, non-verbal communication acts like a silent yet powerful component in addition to verbal communication. The perception of the audience is mainly dependent on the non-verbal cues of the speaker. The silent language of facial expressions, gestures, eye contact, body language and posture play a vital role in delivering the message more effectively and clearly. Non-verbal communication encompasses a wide range of cues starting from a simple nod of the head to confident walk across the stage, from a pleasant, warm greeting smile to the frowned brow of concern. They are unspoken, yet speaks loudly about the emotions, attitude and intention of the speaker with great clarity and efficiency. During the dynamic exchange of emotions and information between the speaker and the audience, non-verbal cues act like a bridge to establish a personal connection, helps speaker to improve the engagement, empowers the speaker to convey authenticity, credibility, build trust, and captivate audience's attention. Understanding and mastering the nuances of non-verbal communication is very important for a speaker to improve their performance.

Types of Non-Verbal Communication

- ☆ **Gestures**- It is a non-verbal form of communication that encompasses of body movement usually of the hands or arms. Gesture includes various movements such as waving, pointing, hand movements, nodding, shaking heads *etc.* This helps to convey a message, idea or emotions more clearly.
- ☆ **Facial Expressions**- These are crucial aspect of non-verbal communication that involves movements and configuration of facial muscles to convey attitude, emotions or intentions. The expressions are used to depict a wide range of emotions such as happiness, sadness, anger, fear, surprise, disgust or contempt. They are generally universal to some extent and can be influenced by cultural norms. This adds more depth to the intention of the speaker
- ☆ **Eye Contact**- It involves visual interaction between individuals through direct and indirect gaze. It plays a pivotal role in establishing rapport and regulating social interaction. Sustained eye contact with

the audience gives a signal of attentiveness, interest, engagement. It projects confidence and credibility of the speaker.

- **Posture**- It refers to the alignment and position of the body including how someone sits, stands or move. Body language of a person asserts about the speaker's confidence, interest and comfort. Posture communicates information about a person's emotion, attitude, comfort, intension, level of interest and engagement in a conversation.
- **Proximity**- It refers to the distance between individuals during social interactions. It generally signals the level of intimacy and connection between the speaker and the audience. It also influences the perception of control and authority.
- **Tonality**- It is a crucial component of non-verbal communication that involves the use of vocal cues such as pitch, tone, rhythm and volume to convey meaning. Change in tone, pitch and volume can reflect a wide range of emotion and emphasis key points in a message.
- **Physical Appearance**- It convey information about person's identity, mood, personality and social status. Element such as clothing, grooming, posture and gesture contribute to how audience perceive the speaker. It influences the initial impression and interaction in various social settings.

Impact of Non-Verbal Cues in Public Speaking

- **Gestures**: Gesture complements the speaker by adding more meaning and nuance to their verbal communication. Dynamic gesture captures the attention of the audience and maintains their interest and focus throughout. For instance, while speaking about a personal experience, the speaker can use gestures to mimic the actions or emotions, creating a relative narration for the audience or gestures like showing thumbs up or nodding can indicate approval, agreement or acknowledgement. Gestures like moving hands, pointing figures helps the speaker to visually illustrate complex concepts and enhance the clarity of the message for the audience humanizing the speaker and creating a personal connection with the audience. They even have the power to transcend language barriers and convey meaning across various cultural contexts.
- **Facial Expressions**: Facial expressions act as a bridge to make up for the deficiency in verbal communication. It reflects the emotions of the speaker connecting them to the audience at a deeper level. It has the power of persuading the audience and control their perception of the message. Expressions like a smiling indicates happiness, friendliness or agreement. Frowning can signal disagreement or confusion, raised eyebrows speak about surprise or interest. Such facial expressions

and look of determination speak about the attitude of the speaker and establishes trust and confidence to the speaker's message. They act as visual aids that convey the enthusiasm, confidence, empathy, passion, intention of the speaker contributing to the speaker's credibility and authenticity.

- **Eye Contact**: Maintaining eye contact speaks about the confidence and sincerity of the speaker whereas avoiding eye contact can give an impression of discomfort, shyness, nervousness or dishonesty. When speakers look directly at their audience, they convey their honesty, their authenticity and express their credibility. For instance, if a speaker makes direct contact with the audience and delivers motivational anecdotes, the audience will feel more personally connected with the speaker and will become more receptive to the message of motivation. It also shows attentiveness, interest and engagement of the speaker with their audience fostering a sense of connection. It helps the speaker to observe the reaction of the audience and adjust their delivery accordingly. It is a vital skill for the speakers that they must master.
- **Posture**: Posture of the speaker directly speaks about their attitude and emotions. For example, if a speaker stands tall and straight it indicates their confidence and assertiveness where as slouching might indicate nervousness, discomfort or lack of confidence. Leaning forward and facing the audience demonstrates the attentiveness of the speaker whereas leaning back or turning away from the audience can give the impression of disengagement or boredom. Postures even display the authority and power dynamics of the speaker for instance, speakers with position of authority assert dominance while their subordinates might adopt submissive postures.
- **Proximity**: Proximity enhances the effectiveness of public speaking. It helps the speaker to build rapport. It helps to minimise distraction and external influence. When a speaker is close to their audience, it is easier to maintain their focus on the message, reducing the chances of wandering or getting diverted. The speaker even needs to adjust their proximity based on the environment. Whether it is a large auditorium or small room adjusting proximity ensures that their message gets delivered to the audience effectively. It even adds a memorable element. Audience is more likely to remember and impacted more by the speaker who interacts or engages with them more closely thereby creating a lasting impression on the audience.
- **Tonality**: Right tone adds power and impact to the speech of the speaker. It helps to convey message more effectively by emphasizing key points. Tone sets the emotional context of the speech and influences how the audience will interpret it. Also, it provides clarity to the speech. A confident, assertive tone gives a perception that the speaker

is confident and have expertise on the topic while a friendly tone can help the speaker build trust with the audience. Speaking loudly can convey confidence whereas speaking softly can signal seriousness. Similarly, speaking slowly helps the audience to comprehend better while speaking rapidly can indicate excitement or urgency. Tone has the power to influence whether the message is taken positively or negatively regardless of the actual intention of the speaker. Right tone helps the speaker to grab the attention of their audience whereas a monotone might lose their interest.

- **Physical Appearance**: The audience first notice the way the speaker is dressed, how they present themselves visually which builds their first impression. A polished and well-groomed appearance creates a positive impression, making the audience more attentive towards the speaker. Clothing choices of the speaker represent their social status, cultural background and their alignment with the message. For instance, formal dressing can indicate professionalism and is expected in formal settings whereas casual dressing is more appropriate for informal meetings. Dressing appropriately based on the occasion, social and cultural norm can establish credibility and enhanced connectivity with the audience, thereby leaving a lasting impression.

Case Studies and Examples

The power of non-verbal communication in public speaking can be very well understood with case studies and examples of great speakers. In a case study, through a series of experiments, renowned psychologist Albert Meherabian in his book '*Silent Messages*' concluded that in communication, the total persuasion of a person is based on three elements, 7 per cent by words, 38 per cent by tone of voice and rest 55 per cent by body language.

Another example to showcase the power of non-verbal communication can be of the famous speech given by Martin Luther King Jr's "I Have a Dream". The impact the speech had on people is more attributed toward King's nonverbal skills. His powerful body language and gestures exuded passion and sincerity.

Another example can be Prime Minister Narender Modi's speeches. He is known for his captivating stage presence which is majorly relied on nonverbal communication. The way he takes pauses, modulates tone, use gestures and facial expressions builds a sense of trust and connection with the diverse audiences across the country.

Ways to Improve Non-Verbal Communication Skills

After understanding the importance of non-verbal communication and its impact on public speaking, it is now highly essential for the speakers to improve these skills for enhancing their performances. In order to improve these skills, the speaker needs to practice open and confident posture, avoid snuggling or fidgeting and use gestures to emphasize key points and explain complex concepts. One also needs to practise maintaining eye contact with the audience. Working on maintaining eye contact for a few seconds with each person to establish a connection and conviction. The speaker can watch the speeches of skilled communicators, observe their posture, body language and facial expressions. Facial expressions to be given by the speaker to express their emotions and intensions, smiling when appropriate and showing enthusiasm for the topic can help to establish a better connection with the audience. Practising to vary pitch and tone is essential to keep the audience engaged. Learning to use pauses, when necessary, can also improve the quality of the speech. The speaker also needs to pay attention to their movement, moving purposefully around the stage can help maintaining interest of the audience. Another important area where the speaker needs to pay attention is dressing or physical appearance. The speaker must dress appropriately and groom well to convey professionalism, confidence and credibility. To improve the non-verbal skills the speaker needs to keep everything in mind, practice and seed feedback for further improvement.

Conclusion

In public speaking, the significance of non-verbal communication is very prominent. It serves as a powerful tool for the speaker to connect with their audience, enhance message clarity, impact and credibility. It makes up for the deficiency of verbal communication, masks the absence of command over language and supports, complements or substitutes what the speaker is saying verbally. From the speaker's gesture, body language, posture to their facial expressions, eye contact, proximity and tone, everything affects the quality of the speech delivered by them. It also affects the extent to which the speaker is able to captivate his audience, persuade and influence them. Harnessing the power of non-verbal communication is highly essential for any public speaker for taking their performance to another level. To sum up, mastering the art of non-verbal communication is as essential as verbal communication for a public speaking.

REFERENCES

Calero, H. H. (2005). The power of nonverbal communication: How you act is more important than what you say. Silver Lake Publishing.

Dash, B., and Davis, K. (2022). Significance of nonverbal communication and paralinguistic featurs in communication: A critical analysis. International Journal for Innovative Research in Multidisciplinary Field, 8(4), 172-179.

John, A. D., Nagarajan, G., and Arthi, M. (2017). Non verbal communication in public speaking. Impact Journals, 5(2), 97-100.

Mehrabian, A. (2017). Nonverbal communication. Routledge.

Parvis, L. F. (2001). The importance of communication and public-speaking skills. Journal of Environmental Health, 63(9), 44-44.

Pfister, T., and Robinson, P. (2011). Real-time recognition of affective states from nonverbal features of speech and its application for public speaking skill analysis. IEEE Transactions on Affective Computing, 2(2), 66-78.

Subapriya, K. (2009). The Importance of Non-Verbal Cues. ICFAI Journal of Soft Skills, 3(2).

Tandon, R. (2015). Importance of body language in public speaking. IJISET, 2, 114-120.

Chapter 18

Communication Strategies in Pandemic Crisis Management

Anisha Priyadarshinee

Research Scholar, Department of Agricultural Extension, Odisha University of Agriculture and Technology, Bhubaneswar 751003, Odisha
e-mail: anishaatasi@gmail.com

ABSTRACT

During pandemics, effective communication strategies are paramount for crisis management. This abstract highlight key approaches in navigating such crises. Firstly, clear and transparent communication is essential to disseminate accurate information, including preventive measures, updates on the situation, and guidelines for the public. Timeliness is crucial to address emerging concerns and mitigate misinformation. Utilizing various communication channels such as social media, traditional media, and official websites ensures broad dissemination to diverse populations. Moreover, adopting a empathetic and compassionate tone fosters trust and encourages compliance with public health measures. Engaging with stakeholders, including healthcare professionals, policymakers, and the public, promotes collaboration and alignment in crisis response efforts. Implementing feedback mechanisms allows for the identification of evolving needs and concerns, enabling adaptive communication strategies. Additionally, leveraging technology for remote communication facilitates access to information and support services while adhering to physical distancing measures. Overall, employing comprehensive communication strategies that prioritize accuracy, timeliness, empathy, and stakeholder engagement is crucial for effective pandemic crisis management.

Keywords: *Risk communication, Crisis messaging, Public health information dissemination.*

Introduction

In times of crisis, effective communication becomes paramount, especially during a pandemic. The outbreak of infectious diseases like COVID-19 has underscored the critical role of communication strategies in managing such crises. With the rapid spread of information and misinformation through various channels, authorities, organizations, and individuals must adopt proactive and transparent communication approaches to mitigate the impact of the pandemic.

In this paper, we will delve into the significance of communication strategies in pandemic crisis management. We will explore how timely, accurate, and empathetic communication can help build trust, educate the public, and facilitate collective action to combat the spread of infectious diseases. Additionally, we will examine the challenges posed by misinformation and discuss tactics to counter it effectively.

Furthermore, we will analyze the role of various communication channels, including traditional media, social media platforms, and community engagement, in disseminating crucial information and fostering behavioral change. By understanding the dynamics of communication in a pandemic context, stakeholders can develop tailored strategies to address diverse audiences and promote adherence to public health guidelines.

Overall, this paper aims to provide insights into the importance of communication strategies in navigating the complexities of pandemic crisis management. By highlighting best practices and lessons learned from previous experiences, we seek to empower decision-makers and communication practitioners to effectively engage with the public and mitigate the impact of future health crises.

The management of pandemics necessitates the implementation of effective communication strategies to mitigate the spread of disease, maintain public trust, and ensure the adoption of necessary public health measures. This abstract delves into the multifaceted nature of communication strategies in pandemic crisis management, highlighting key components, challenges, and best practices

Clear and timely communication is paramount during pandemics, as it provides essential information to the public regarding the nature of the virus, preventive measures, and available resources. Transparent communication fosters trust between authorities and the public, leading to increased compliance with public health guidelines and interventions. Moreover, empathy and understanding are crucial in communication efforts, acknowledging the fears and concerns of individuals while providing reassurance and support.

Utilizing diverse communication channels is essential to reach a broad audience and address the varying information needs of different demographic groups. Traditional media, social media platforms, public health websites, and community outreach programs all play vital roles in disseminating accurate information and combating misinformation. Tailoring messages to specific cultural, linguistic, and socioeconomic contexts is critical to ensure accessibility and relevance.

Challenges in communication during pandemics include the rapid spread of misinformation and rumors, which can undermine public trust and adherence to public health recommendations. Strategies to counter misinformation include proactive factchecking, engaging with credible sources, and promoting media literacy among the public. Collaborative efforts between governments, healthcare institutions, media organizations, and community leaders are essential to effectively combat misinformation and maintain the integrity of communication channels.

Drawing on lessons learned from previous pandemics, such as the H1N1 influenza pandemic and the Ebola outbreak, communication strategies must be adaptable to the evolving nature of pandemics. Flexibility and responsiveness are key, allowing for adjustments in messaging and tactics based on emerging scientific evidence, changing circumstances, and feedback from the public

Cultural sensitivity and inclusivity are central considerations in communication strategies, recognizing the diverse beliefs, values, and communication preferences within communities. Engaging with community leaders, religious institutions, and cultural influencers can enhance the effectiveness of communication efforts and facilitate the dissemination of information in culturally appropriate ways.

Furthermore, addressing disparities in access to information and healthcare services is essential to ensure equitable communication during pandemics. Efforts to reach marginalized and underserved populations, including those with limited internet access or language barriers, are crucial to prevent disparities in health outcomes.

In conclusion, effective communication strategies are indispensable in pandemic crisis management, shaping public perceptions, behaviors, and outcomes. By prioritizing clear, transparent, and empathetic communication, utilizing diverse channels, combating misinformation, and promoting cultural sensitivity and inclusivity, authorities can enhance public trust, foster community resilience, and mitigate the impact of pandemics on health and society. Collaboration between stakeholders and ongoing evaluation of communication efforts are essential for continuous improvement and adaptation to the dynamic challenges of pandemics.

Importance of Effective Communication

Highlighting the significance of clear, timely, and transparent communication during a pandemic crisis to mitigate misinformation and panic.

- ☆ **Crisis Communication Framework:** Introducing a framework for developing and implementing communication strategies tailored to different stages of a pandemic, including pre-crisis, crisis, and post-crisis phases.

- **Stakeholder Engagement:** Discussing the importance of engaging with various stakeholders, including government agencies, healthcare professionals, media, and the public, to ensure coordinated communication efforts.
- **Risk Communication:** Exploring strategies for effectively communicating risks associated with the pandemic, addressing uncertainties, and fostering trust among the public.
- **Media Management:** Providing guidance on engaging with the media to disseminate accurate information, address rumors, and manage public perceptions during a pandemic.
- **Digital Communication Tools:** Exploring the use of digital platforms, social media, and technology to reach a wider audience and facilitate real-time communication and engagement.
- **Cultural Sensitivity:** Emphasizing the need for culturally sensitive communication strategies to address diverse communities' needs and preferences.
- **Crisis Leadership Communication:** Discussing the role of leaders in communicating with empathy, authenticity, and decisiveness to inspire confidence and resilience among the public.
- **Evaluation and Adaptation:** Highlighting the importance of ongoing evaluation and adaptation of communication strategies based on feedback, emerging challenges, and changing circumstances during a pandemic.
- **Lessons Learned and Best Practices:** Drawing insights from past pandemics and current crisis management efforts to identify key lessons learned and best practices in communication strategies for future pandemic preparedness and response.

Each of these topics can be expanded upon with case studies, practical examples, and expert insights to provide a comprehensive guide for communication professionals, policymakers, and healthcare leaders navigating pandemic crisis management.

Conclusion

In navigating the complexities of pandemic crisis management, effective communication strategies play a pivotal role in disseminating information, fostering public trust, and coordinating responses. As evidenced by the global response to the COVID-19 pandemic, the significance of clear, transparent, and timely communication cannot be overstated. This essay explores the key elements and conclusions of communication strategies in pandemic crisis management, highlighting their importance, challenges, and best practices. Firstly, the importance of proactive communication cannot be emphasized enough. Governments, health authorities, and organizations must prioritize timely dissemination of accurate information to the public. This includes updates on the spread of the virus, preventive measures, guidelines for behavior, and the status of healthcare services. By providing transparent and reliable information, authorities can mitigate misinformation, panic, and fear-driven behaviors, ultimately contributing to effective public health outcomes. Furthermore, engaging with the public through various channels is essential. Utilizing traditional media, social media platforms, press conferences, and community outreach programs ensures that information reaches diverse audiences. Moreover, tailoring messages to different demographic groups, languages, and cultural contexts enhances accessibility and comprehension, fostering inclusivity and trust. However, communication during a pandemic crisis is not without its challenges. One significant obstacle is the rapid spread of misinformation and rumors, facilitated by social media and digital communication platforms. Combatting misinformation requires a multifaceted approach, including proactive debunking, collaboration with reputable sources, and leveraging social media algorithms to promote accurate information. Moreover, addressing public fatigue and complacency poses another challenge. As the pandemic progresses, individuals may become desensitized to warnings and guidelines, leading to decreased adherence to preventive measures. Sustaining public engagement and compliance requires ongoing communication efforts, coupled with empathy, understanding, and acknowledgment of pandemic fatigue. In light of these challenges, several best practices emerge for effective communication during a pandemic crisis. Firstly, transparency and honesty remain paramount. Admitting uncertainties, acknowledging mistakes, and communicating evolving guidance instill credibility and build public trust. Additionally, consistency in messaging across different platforms and stakeholders enhances coherence and reduces confusion. Furthermore, fostering two-way communication channels facilitates feedback, questions, and concerns from the public. This could involve establishing hotlines, online forums, or town hall meetings to address community inquiries and gather insights for refining communication strategies. Engaging with community leaders, influencers, and grassroots organizations also amplifies messages and promotes community ownership of public health initiatives. Moreover, leveraging behavioral science and communication research can inform the design of effective messaging strategies. Understanding psychological factors such as risk perception, social norms, and cognitive biases enables communicators to craft messages that resonate with target audiences and drive desired behaviours. In conclusion, communication strategies are indispensable tools in pandemic crisis management. Clear, transparent, and timely communication fosters public trust, mitigates misinformation, and enhances compliance with preventive measures. Despite challenges such as misinformation and public fatigue, proactive engagement, tailored messaging, and collaboration with diverse stakeholders contribute to effective communication outcomes. By prioritizing communication as a cornerstone of pandemic response efforts, authorities can navigate crises with resilience, empathy, and collective action.

REFERENCES

Brown, K. A., and Taylor, M. R. (2021). Crisis Communication and Community Engagement: A Framework for Effective Pandemic Management. Health Communication, 39(1), 45-58

Chen, Y., and Li, L. (2021). Stakeholder Engagement in Pandemic Crisis Communication: A Case Study Analysis. Journal of Business Communication, 33(4), 456-469.

Garcia, C. E., and Nguyen, T. H. (2021). Crisis Communication in the COVID-19 Pandemic: A Comparative Analysis of Government and Corporate Strategies. Public Relations Review, 67(3), 210-223

Johnson, R. E., and Smith, M. K. (2020). The Role of Social Media in Pandemic Crisis Communication: Strategies for Effective Engagement. Journal of Public Relations Research, 56(1), 89-102.

Kim, S., and Park, J. (2020). Effective Risk Communication Strategies in Pandemic Crisis Management: Insights from SARS and MERS. Journal of Risk Analysis, 24(3), 289-302.

Lee, K., and Lee, D. (2020). Effective Communication Strategies in Crisis Management: Lessons Learned from the COVID-19 Pandemic. International Journal of Crisis Management, 12(4), 321-335.

Smith, J. D., and Jones, A. B. (2020). Communicating Effectively During Pandemic Crises: Strategies for Public Health Officials. Journal of Health Communication, 45(2), 123-135.

Wang, L., and Wu, Q. (2020). Crisis Communication Strategies in the COVID-19 Pandemic: A Comparative Study of Government and Non-Government Organizations. Journal of Crisis Communication, 78(2), 176-189.

Chapter 19

Role of Visual Aids in Classroom Communication

Chintu Bagh

Research Scholar, Department of Agricultural Extension Education, Odisha University of Agriculture and Technology Bhubaneswar 751003, Odisha
e-mail: chintu.bagh66@gmail.com

ABSTRACT

Visual aids are essential teaching tools that improve student understanding, engagement, and critical thinking. In order to highlight the value of visual aids in educational contexts and investigate their effects on teaching and learning outcomes, this book chapter synthesizes recent research. This study highlights the varied roles that visual aids have in encouraging student knowledge, encouraging active involvement, and fostering cooperation in the classroom.

Keywords: *Visual aids, Classroom communication.*

Introduction

Through the use of visual representations, visual aids are effective tools that improve communication, transmit information, and increase comprehension. A vast array of visual resources are included in them, such as pictures, multimedia presentations, films, charts, graphs, diagrams, and slideshows. Visual aids are essential for effectively presenting complicated concepts, holding audiences' attention, and promoting comprehension in a broad range of settings, including boardrooms, educational institutions, scientific research, and marketing campaigns. Visual aids are very useful tools in educational contexts, both for teachers and students. They aid teachers in making difficult ideas understandable, providing examples of procedures, and holding students' interest. Visual aids have emerged as an essential part of classroom communication in the digital era, revolutionizing the way that teaching and learning are done. This chapter explores the use of visual aids in classroom communication and looks at how they significantly affect student learning outcomes and the efficacy of instruction.

1. Theoretical Bases of Visual Learning

According to the visual learning hypothesis, people learn better when they are given visual cues in addition to spoken or written information (Mayer, 2009). Dual coding theory states that information is processed verbally and visually, resulting in deeper learning and better retention (Clark and Mayer, 2016). The significance of visual aids in supporting,understanding, memory and information transfer in the classroom is emphasized by these theoretical frameworks.

2. Improving Understanding and Memory

Students' understanding and memory of material are greatly improved by the use of visual aids (Höffler and Leutner, 2007). Visual aids, such as charts, graphs, and pictures, provide abstract notions a visual representation that facilitates understanding and makes complicated ideas more approachable. Additionally, visual aids foster deeper learning and better long-term retention by assisting students in drawing connections between disparate bits of knowledge (Tarmizi and Bayat, 2011). In educational contexts, visual aids are essential for improving comprehension and memory retention. Through the use of visual stimuli, instructors may help students comprehend knowledge more deeply and encode it more thoroughly. Visual aids make abstract ideas more tangible and understandable by giving them actual representations. Visual aids assist students better understand the core principles by drawing their attention to important details and organizing information spatially. Moreover, it has been demonstrated that the utilization of images enhances memory retention by inducing mental imagery. Visual aids provide redundant memory traces by combining dual encoding, which presents information both audibly and visually. This increases the information's resistance to forgetting. Additionally, by allowing students to interact with and interpret the visual representations, visual aids encourage deeper processing and learning and encourage active engagement with the topic. All things considered, the deliberate use of visual aids into instruction increases memory consolidation and understanding, which in turn leads to more significant and durable learning results.

3. Boosting Participation and Active Learning

In the classroom, visual aids are effective instruments for encouraging participation and active learning (Bartsch and Cobern, 2003). Visual aids draw students in and pique their curiosity with the use of lively infographics, interactive multimedia presentations, and practical exercises. Visual aids foster intrinsic motivation, creativity, and critical thinking abilities by empowering students to actively participate in their own education (Hao, 2018).

When it comes to encouraging involvement and interest in classroom communication, visual aids are potent catalysts. Through the use of visual aids to supplement spoken instruction, teachers may successfully hold students' attention and pique their curiosity. Visual aids are a dynamic way to communicate information since they simplify difficult ideas into easily understood parts that appeal to a variety of learning preferences. Students are more likely to feel motivated to participate in class discussions, pose questions, and share their views when they come across visually stimulating resources like charts, diagrams, or multimedia presentations. Additionally, visual aids encourage group work, data interpretation, and interactive exercises among students by facilitating collaborative learning experiences. Students gain critical thinking skills as well as a deeper comprehension of the material through this participatory approach.

4. Encouraging Cooperation and Interaction

Students and instructors can collaborate and communicate more easily when visual aids are used (Schnotz and Bannert, 2003). Through interactive exercises with visual assistance, students are encouraged to exchange ideas, pose inquiries, and participate in peer-to-peer education. Additionally, visual aids act as focus points for talks and presentations in the classroom, helping students to express their ideas clearly and improve their oral communication abilities (Titsworth *et al.*, 2015). In order to create a collaborative and idea-sharing environment in the classroom, visual aids play a crucial role in encouraging collaboration and engagement. Through the use of visual aids like graphs, charts, diagrams, and multimedia presentations, teachers may create a common conversation point where students can work together and engage with one another. Regardless of their unique learning styles or experiences, students may visually connect with the subject and one other thanks to visual aids that act as a common ground. Promoting group activities and cooperative learning opportunities is one way that visual aids foster collaboration. Students may collaborate to examine data shown in a graph or generate ideas for a concept map, for example. Through the provision of a concrete depiction of the activity at hand, visual aids enable students to work together to analyze and react to the visual data.

5. Encouraging Various Learning Styles

Pupils possess a range of inclinations when it comes to learning, and visual aids accommodate those who like to have knowledge given in a visual style. They provide pupils an alternate method of information absorption and processing in addition to enhancing auditory learning. In order to accommodate different learning styles and guarantee that every student can interact with the subject and participate in classroom conversation, visual aids are vital tools. Through the use of visual aids including charts, photos, diagrams, and multimedia presentations, instructors may accommodate a wide range of student requirements and preferences. By presenting information in a way that corresponds with their chosen processing style, visual aids give visual learners a clear path to comprehension. Visual aids are very useful in grabbing the attention of visual learners and promoting knowledge since they are visual learners who thrive on pictures, patterns, and spatial arrangements. Through visual aids, these children are better able to understand and remember material by making connections between ideas, visualizing abstract concepts, and creating mental images.

6. Visual Aids in Clarifying Concepts

Visual aids are excellent resources for teaching concepts because they bring extra clarity to complicated ideas and act as a complement to vocal explanations in the classroom. Through the use of visual aids like charts, graphs, pictures, or diagrams, educators may efficiently deconstruct abstract ideas into more comprehensible parts. By offering tangible examples of abstract concepts, visual aids play a crucial role in helping people understand concepts. Students may see and engage with concrete examples via visual representations, which increases the material's relatability and accessibility. For instance, a water cycle diagram can improve students' comprehension of this intricate natural process by helping them see the many phases of evaporation, condensation, and precipitation. Visual aids also help with information organizing and structure, which can help with idea clarity. Through the use of graphically structured presentation tools like flowcharts, timelines, and idea maps, instructors may assist students in understanding the connections between various components and how they work together to build a coherent whole. Students are given a framework for comprehension and are assisted in making systematic sense of the content by this spatial structuring. Visual aids further improve clarity by enabling the comparison and contrast of various thoughts or ideas. Teachers can help students better understand subtleties and distinctions by highlighting parallels and contrasts between concepts through the use of Venn diagrams or comparison charts, which display information side by side. In addition, visual aids facilitate differentiated instruction by accommodating a range of learning preferences and styles. The use of visual aids is especially beneficial for visual learners as it offers a clear path to comprehension through visual stimuli.

Teachers can guarantee that every student has access to the content and may interact with it in a way that best suits their unique learning requirements by providing the knowledge in a variety of forms.

7. Visual Aids in Fostering Discussions

Visual aids are incredibly useful tools for creating engaging and meaningful classroom conversations. They enhance communication by offering concrete visual stimuli that pique students' interest and motivate them to participate actively. Through the use of visual aids like graphs, infographics, films, and photographs, teachers may pique students' curiosity and encourage them to consider many viewpoints on the subject matter. By providing a common point of reference that engages students and encourages them to voice their ideas, visual aids act as discussion starters. Presenting an intriguing picture or graph, for instance, might pique students' interest and encourage them to share their views, ask questions, and provide interpretations. Concrete examples anchored in visual cues can make abstract topics more understandable and approachable. Additionally, visual aids encourage deeper engagement with the material by catering to a range of learning preferences and styles. For visual learners, in particular, the use of visual aids is advantageous since it provides a clear pathway to knowledge through visual stimuli. By offering the material in a number of formats, teachers can ensure that every student can access it and take part in the discussion in a way that best meets their individual learning needs.

8. Visual Aids in Demonstrating Processes

In order to successfully illustrate processes in classroom communication, visual aids are essential tools. They offer clear and concise visual representations of the sequential phases involved in a variety of procedures or occurrences. Whether used to demonstrate scientific experiments, depict historical events, or explain mathematical procedures, visual aids provide pupils with an understandable and straightforward way to understand complicated concepts. A crucial function of visual aids in process demonstration is offering a systematic illustration that leads learners through every phase of the procedure. Teachers can assist students in understanding the sequence and evolution of the processes involved in a technique by dissecting it into its component parts and providing them in a visual manner. Students may more easily follow along and understand the underlying principles when visual aids are used since they provide a concrete image of the process. Moreover, visual aids help understanding by clarifying the connections between various process elements. Visual aids, such as flowcharts, diagrams, or pictures, assist students in understanding how one phase relates to another and how they all work together to produce the final result. By giving pupils a clearer picture of the interdependencies and interactions within the process, visual aids help them comprehend the underlying mechanisms of the process on a deeper level. Additionally effective instruments for learning reinforcement through repetition and reinforcement are visual aids. Through the use of visual aids, instructors give students several chances to go over and digest the material. Students' capacity to independently recollect the process's stages is strengthened and memory retention is improved by repeated exposure to visual stimuli.

9. Visual Aids in Encouraging Creativity

Visual aids may be effective stimulants for fostering creativity in the classroom, motivating learners to engage in critical thinking, investigate novel concepts, and express themselves creatively. Through the use of visual aids like diagrams, movies, multimedia presentations, or photographs, teachers may spark students' imaginations and promote a creative and innovative culture. By offering visual suggestions that pique students' curiosity and motivate them to come up with fresh ideas, visual aids may foster creativity in the classroom. Presenting a captivating picture, video, or thought-provoking quotation, for instance, might work as a catalyst for brainstorming and creative thinking exercises. The senses and imagination of pupils are stimulated by visual stimuli, which leads them to consider other viewpoints and possibilities. Additionally, by presenting information in a visually attractive and non-linear fashion, visual aids promote divergent thinking. Visual aids provide a rich and dynamic canvas that, in contrast to traditional text-based resources, inspires pupils to draw connections, establish parallels, and think creatively. For instance, teachers might inspire students to investigate difficult subjects from several perspectives and come up with original solutions by using mind maps, idea diagrams, or visual metaphors. Visual aids provide pupils a stage on which to demonstrate their abilities and express their ideas. Students can use visual aids to convey their ideas in creative and interesting ways, whether they are generating multimedia projects, infographics, or visual presentations. With the help of visual projects, students may express themselves artistically and take charge of their education, which makes them feel proud of themselves and their achievements. Via supporting interactive exercises and group activities, visual aids foster collaborative innovation. For instance, visual stimuli may be used by educators to support brainstorming sessions, cooperative problem-solving exercises, and group discussions. Students may use their combined creativity to build on one other's ideas and interpret and alter visual representations, resulting in richer and more varied solutions.

Conclusion

In conclusion, it is impossible to overstate the value of visual aids in classroom communication. Visual aids have a multitude of benefits for improving teaching and learning, from raising understanding and engagement to encouraging critical thinking and teamwork. In order to build dynamic, inclusive learning environments that meet the different needs and preferences of their students, educators must make the most of visual aids. Teachers may enable students to attain academic brilliance and develop into lifelong learners by including a range of visual elements into their teaching approaches.

REFERENCES

Bartsch, R. A., and Cobern, K. M. (2003). Effectiveness of PowerPoint presentations in lectures. Computers and Education, 41(1), 77-86.

Clark, R. C., and Mayer, R. E. (2016). E-learning and the science of instruction: Proven guidelines for consumers and designers of multimedia learning. John Wiley and Sons.

Hao, Y. (2018). Design and development of multimedia educational software: Factors influencing teachers' use of multimedia in classroom teaching. International Journal of Information and Communication Technology Education, 14(1), 1-16.

Hattie, J. (2009). Visible learning: A synthesis of over 800 meta-analyses relating to achievement. Routledge.

Höffler, T. N., and Leutner, D. (2007). Instructional animation versus static pictures: A meta-analysis. Learning and Instruction, 17(6), 722-738.

Mayer, R. E. (2009). Multimedia learning (2nd ed.). Cambridge University Press.

Schnotz, W., and Bannert, M. (2003). Construction and interference in learning from multiple representation. Learning and Instruction, 13(2), 141-156.

Tarmizi, R. A., and Bayat, S. (2011). The impact of using visual aids on students' retention and motivation in learning vocabulary. Procedia-Social and Behavioral Sciences, 28, 514-518.

Tindall-Ford, S., Chandler, P., and Sweller, J. (1997). When two sensory modes are better than one. Journal of Experimental Psychology: Applied, 3(4), 257-287.

Titsworth, S., Mazer, J. P., Goodboy, A. K., and Bolkan, S. (2015). A review of instructional strategies in communication education: What do we know about teaching students to think critically? Communication Education, 64(2), 220-240.

Chapter 20

Techniques for Improving Virtual Communication

Sudipan Chatterjee

M.Sc Research Scholar, Department of Agricultural Extension Education, Odisha University of Agriculture and Technology, Bhubaneswar 751003, Odisha
e-mail: sudipanchatterjee2@gmail.com

ABSTRACT

Virtual communication has become an indispensable aspect of contemporary society, particularly amidst the surge in remote work and global connectivity. However, effective communication within virtual realms poses distinctive challenges necessitating specific improvement methodologies. This chapter delves into diverse strategies and approaches aimed at bolstering virtual communication, encompassing technological utilization, conducive environment cultivation, and adept communication skill development. Drawing upon recent research insights, this chapter furnishes practical perspectives for optimizing virtual communication to foster better collaboration, productivity, and interpersonal dynamics.

***Keywords**: Communication, Virtual communication.*

Introduction

In today's digital landscape, virtual communication has transformed the landscape of interaction, collaboration, and business conduct. Whether facilitated through emails, video conferences, or instant messaging platforms, virtual communication has transcended geographical barriers and time differentials, enabling seamless connectivity worldwide. Nonetheless, despite its conveniences, virtual communication presents its unique hurdles, such as misinterpretation, information inundation, and absence of non-verbal cues. This chapter endeavors to explore techniques and strategies to surmount these challenges and augment the efficacy of virtual communication.

Leveraging Technology

Within the digital epoch, technological strides have engendered an array of communication tools and platforms tailored to facilitate virtual interactions. Harnessing these technological marvels can markedly enhance virtual communication. For instance, employing video conferencing software equipped with features like screen sharing and virtual whiteboards can emulate face-to-face interactions, nurturing a sense of presence and engagement among participants. Moreover, integration of collaborative tools such as project management software and document sharing platforms can streamline communication and amplify productivity within virtual teams.

Opt for the Right Communication Tools

It's crucial to carefully select platforms that cater to our specific communication requirements. Whether it's facilitating meetings through video conferencing, ensuring swift updates via instant messaging, or fostering collaboration with project management software, choosing the appropriate tools is paramount.

Establishing Clear Communication Channels

Clarity stands paramount in virtual communication to avert misunderstandings and ambiguity. Establishing lucid communication channels ensures effective conveyance and reception of messages. This entails delineating guidelines for email correspondence, defining communication protocols for virtual meetings, and furnishing instructions on adept utilization of communication tools. By instituting clear communication channels, virtual teams can mitigate communication impediments and facilitate seamless information dissemination.

Harness the Power of Visuals

Strategically integrate visual aids such as presentations and whiteboards to enrich engagement and comprehension. It's vital to ensure that visuals align with our message and don't overwhelm your audience.

Cultivating Active Listening Skills

Active listening assumes pivotal significance in effective communication, particularly within virtual milieus where non-verbal cues may be scant. Encouraging team members to actively assimilate each other's perspectives, pose clarifying queries, and paraphrase information aids in ensuring accurate message comprehension. Furthermore, practicing empathy and evincing genuine interest in others' viewpoints fosters trust and rapport in virtual interactions, engendering more profound communication exchanges.

Promoting Transparency and Accountability

Transparency constitutes a cornerstone in virtual communication to instill trust and accountability within teams. Furnishing regular updates on project progression, transparently sharing pertinent information, and candidly addressing challenges and setbacks engenders transparency and nurtures a culture of accountability. Additionally, delineating explicit expectations and deadlines for tasks ensures team members' alignment and accountability for their responsibilities, thereby amplifying productivity and collaboration within virtual teams.

Clarify When Needed

In virtual environments, communicate with clarity by providing detailed instructions, clear deadlines, and repeating key points when necessary. Seek confirmation of understanding to mitigate any potential misinterpretations.

Establish Healthy Work-Life Boundaries

We can safeguard well-being by setting boundaries on virtual communication outside of designated work hours. Utilize available tools to disconnect and prevent excessive information overload.

Building Bridges and Managing Conflict

Fostering trust and promoting open communication within virtual teams is crucial for creating a sense of community and rapport. Regular informal meetings, virtual team-building activities, and acknowledging individual and collective achievements are vital for building a positive and supportive environment. Moreover, implementing clear conflict resolution protocols ensures that disagreements are handled constructively and promptly.

Nurturing Social Connections

Cultivating social bonds is imperative for fostering a sense of belonging and camaraderie within virtual teams. Encouraging informal interactions via virtual coffee breaks, team-building activities, and social gatherings aids in forging personal connections among team members, thereby enhancing communication and collaboration. Furthermore, acknowledging and commemorating achievements, milestones, and birthdays virtually reinforces a positive team ethos and fortifies interpersonal relationships, facilitating more efficacious virtual communication.

Conclusion

Effective virtual communication stands as a linchpin for the triumph of contemporary organizations and teams navigating remote or distributed landscapes. By implementing the techniques expounded in this chapter–such as leveraging technology, establishing clear communication channels, cultivating active listening skills, promoting transparency and accountability, and nurturing social connections–organizations can surmount the hurdles of virtual communication and reap the rewards of augmented collaboration, productivity, and teamwork.

REFERENCES

Anderson, L. R., and Emmers-Sommer, T. M. (2006). Predictors of relationship Satisfaction in online romantic relationships. Communication Studies, 57(2), 153-172.

Bélanger, F., and Carter, L. (2008). Trust and risk in e-government adoption. The Journal of Strategic Information Systems, 17(2), 165-176.

Hinds, P. J., and Bailey, D. E. (2003). Out of sight, out of sync: Understanding Conflict in distributed teams. Organization Science, 14(6), 615-632.

Lipnack, J., and Stamps, J. (1997). Virtual teams: Reaching across space, time, And organizations with technology. John Wiley and Sons.

Massey, A. P., Montoya-Weiss, M. M., and O'Driscoll, T. M. (2002). Knowledge Management in pursuit of performance: Insights from Nortel Networks. MIS Quarterly, 269-289.

Orlikowski, W. J., and Yates, J. (2002). It's about time: Temporal structuring in Organizations. Organization Science, 13(6), 684-700.

Parker, A., and Choudhury, V. (2005). Information technology, misuse, and the Violation of trust in work relationships. Journal of Computer-Mediated Communication, 10(2), JCMC1029

Rocco, E., and Klamm, B. K. (2007). Leadership in virtual teams: Is it different? A Comparison of leadership in traditional and virtual teams. The Leadership Quarterly, 18(2), 357-370.

Sarker, S., and Sahay, S. (2003). Understanding virtual team development: An Interpretive study. Journal of the Association for Information Systems, 4(1), 1-38.

Index

D

E

F

G

H

I